CLARITY

The Ledger Principle in an Editable World

Lisa J. Scott

This is a work of nonfiction. Certain names, identifying details, and circumstances have been changed to protect privacy. Any resemblance to persons or events beyond those described is coincidental.

ISBNs

Paperback ISBN: 978-1-971735-09-2
Hardcover ISBN: 978-1-971735-08-5
E-book ISBN: 978-1-971735-10-8

Published by
Lisa J. Scott Publishing
Dallas, TX
www.lisajscottpublishing.com

Printed in the United States of America

To the world—
To every soul seeking truth in the noise,
To every heart longing to be met, not managed,
To every mind refusing to be edited,
This book is for you.
May clarity be your anchor, your compass, and your freedom.

Contents

Act IV — The Reckoning

Act V — The Era of Sovereignty

Preface

I didn't set out to write a book about technology.

I set out to write a book about what happens when your inner life meets a world that can edit you.

For a long time, I lived inside what I now think of as *managed reality*—the subtle, civilized version of control. Nothing dramatic. Nothing that would make a good headline. Just a repeating pattern where the world asked me to become more reasonable, more flexible, more quiet, more grateful—until I could no longer tell whether I was living my life or negotiating it.

There were the money meetings—those stiff, weekly rituals that pretended to be about responsibility but were really about leverage. There were interruptions that looked harmless on the surface—small detours, the dog, the phone, the sudden shift away from whatever I was excited to say—until I realized how consistently my voice was being reduced. There were the moments when I found myself explaining my own experience as if it needed a defense. And there was the worst part: I got good at it. I became fluent in minimizing. I paid for peace by making myself smaller.

I thought this was just "relationships."

I didn't yet understand that this is also how systems work.

Because I have spent my career in the messy middle—where the transaction ends and the reality begins. In M&A transitions, there is a moment when the story stops working. The deck can be beautiful, the intent can be sincere, the synergy can be declared—but the actual world still has to function. Contracts have to move. People have to align. Decisions have to be made under pressure. And when they aren't made, when no one owns the record, when accountability is scattered, the same thing always happens:

Reality gets rewritten.

Not by malice. By default.

In those environments, I learned something I didn't yet have language for: the person who keeps the ledger becomes the anchor. The one who can show what's true, what's owed, what changed, what was decided, and what is still open—becomes the difference between chaos and coherence.

That is where this book began.

Not in theory.

In the lived need for a record that holds.

THE WORLD IS BECOMING EDITABLE

Then, quietly, the forms stopped coming.

I didn't mean only the physical paperwork—though that was part of it. I meant the visible structure. The obvious gate. The clear friction that told you where the system was.

Forms were annoying, but they were honest. They showed you what the institution wanted. They showed you the price of entry. They showed you the terms.

When the forms disappeared, I assumed we were becoming freer.

But we didn't become freer.

We became less able to see.

The forms didn't disappear. They became invisible: databases, scoring, automation, identity checks, risk models, silent flags. The new form is a data record. The new gatekeeper is an algorithm. And the most disorienting part is that the system can now make decisions without ever offering you a human face to meet.

At the same time, language itself changed.

Words became cheap.

AI can now produce polished, persuasive, emotionally calibrated language at scale. It can generate convincing explanations, synthetic consensus, perfectly shaped narratives—without any cost behind them. When words become abundant, *trust-by-default* collapses. Not because people become evil, but because persuasion becomes effortless.

In that kind of environment, the old signals break.

Tone stops meaning safety.
Credentials stop meaning truth.
Consensus stops meaning reality.
Performance stops meaning integrity.

So what remains?

Work. Record. Coherence.

Costly signal.

CLARITY IS THE MOST VALUABLE ASSET

This is the central claim of this book:

Clarity is the most valuable asset in an editable world.

Not "clarity" as a vibe.

Not clarity as certainty.

Clarity as a practiced ability to:

- keep your witness intact
- see what is being counted

- notice what is being edited
- understand incentives
- verify before believing
- choose dependencies on purpose
- protect attention like wealth
- stay coherent under pressure

Clarity is not a personality trait. It's a discipline.

And like any discipline, it produces compounding returns.

When you have clarity, you stop negotiating your own reality.

You stop explaining yourself into smaller shapes.

You stop mistaking management for love.

You stop mistaking institutions for parents.

You stop calling confusion "normal."

You begin to live from a clean ledger.

This is not a manifesto

I'm not writing to tell you what to believe politically.

I'm not writing to sell you fear.

I'm not writing to give you an identity to join.

This book is not about being anti-anything.

It's about being oriented.

Some technologies decentralize.

Some centralize.

Some do both at once.

We're going to live through a tug-of-war between decentralizing base layers and increasingly controlled rails—between systems that verify and systems that govern by permission. You can have your own opinions about where that goes. My point is simpler:

No matter what happens out there, you need an inner record that cannot be edited by the room.

That's sovereignty.

Not as ideology.

As lived orientation.

HOW THIS BOOK IS BUILT

This book is structured in acts, because clarity does not arrive as an idea. It arrives as a story.

You'll move through:

- the noise I called normal
- the subtle shrink of being managed, not met
- the cost of being "reasonable"
- the mirror breaking when the old contract fails
- the moment I stopped explaining
- the ledger appearing as something explicit
- the technology turn—AI, data, synthetic signal
- the end of trust-by-default
- the new scarcity: coherence
- proof of work in human systems
- money as a ledger

- institutions as machines
- love, parenting, pressure, delusion
- and finally, the ledger principle—what holds, and how to live it

Between those acts are short tech interludes. They're not written for experts. They're written for humans—people who can feel the world changing and want language for what they already sense.

A NOTE TO THE READER

If you've ever felt yourself getting edited—by a relationship, by a workplace, by a system, by the quiet pressure to be "reasonable"—this book is for you.

If you've ever realized that your life got smaller while you were being praised for being mature, this book is for you.

If you're raising a child who sees too much and doesn't fit neatly into other people's forms, this book is for you.

If you sense that the future will reward performance and punish coherence—and you refuse to become a performance—this book is for you.

This is not a book about control.

It's a book about what cannot be controlled by others once you decide to keep it intact.

Your witness.

Your record.

Your coherence.

Your clarity.

Because clarity isn't just a tool for understanding the world.

It is the asset that protects your life from being rewritten.

ACT I

THE ERA OF PERMISSION

1.

The Noise I Called Normal

I used to think the noise was just life.

The constant hum—emails, meetings, obligations, errands, messages, plans, calendar pings, the low-grade anxiety of keeping everything moving—felt like the price of being a functional adult. The noise was everywhere and it was oddly moralized: if you were busy, you were responsible; if you were overwhelmed, you were important; if you were tired, you were doing it right.

I didn't question it because everyone around me seemed to be living the same way. The noise wasn't an interruption. It was the environment. Like air.

And because it was everywhere, I confused it with reality.

I confused motion with progress.
I confused compliance with safety.
I confused being "reasonable" with being free.

For a long time, I believed the world rewarded people who adapted.

If you could shape-shift—be agreeable at dinner, sharp in a meeting, calm in conflict, productive under pressure—you could survive almost anything. You could keep the peace. You could keep the job. You could keep the relationship. You could keep the family story intact.

I was good at adapting.

That's one reason it took me so long to notice that adaptation can become a kind of disappearance.

The quiet bargain

The noise I called normal wasn't just schedule noise. It was relational noise.

It was the subtle negotiation happening under ordinary conversation.

A normal call with my mother could flow for a while—updates, surface life, mild laughter—and then, predictably, something would shift. If the subject moved toward something I was excited about or something that was *mine*, the atmosphere would change. Sometimes it was a sudden interruption, sometimes it was a comment to the dog, sometimes it was a hard pivot to something else.

It wasn't dramatic enough to accuse. That was the genius of it.

It was just enough to redirect the energy.

So I learned to do what so many people learn to do: I stopped.

I didn't push the moment. I didn't insist on being heard. I didn't demand attention. I simply let my sentence trail off and watched the conversation become about something else.

From the outside, that looks like maturity. Inside, it feels like vanishing.

I didn't call it vanishing then. I called it being polite. I called it not making a big deal. I called it not being needy.

But the body keeps its own record. And if you're honest, you can feel the moment your own aliveness is subtly unwelcome.

That's the first kind of noise: the noise that teaches you to edit yourself.

MONEY AS A KIND OF STATIC

Then there was money noise.

I lived inside a version of adulthood where money was always present as a subtle pressure—like static in the walls. It wasn't only about paying bills. It was about the emotional architecture around money: who gets to decide, who gets to question, who gets to monitor.

The weekly budget meetings were framed as responsibility, as teamwork, as good marriage hygiene. But they never felt neutral. They felt one-sided. They carried an unspoken message: *you are being managed.*

Even though I made more salary.

Even though I was capable and conscientious.

Even though I wasn't reckless.

That's the strange thing about relational power: it doesn't always attach to facts. It attaches to posture. It attaches to the ritual.

And those meetings were a ritual.

They taught my nervous system something I didn't want to admit:

Money wasn't just money.
Money was permission.
Money was scrutiny.
Money was leverage disguised as care.

I lived with that static long enough that it started to feel normal too.

The performance of gratitude

There were other small moments—small enough to dismiss, sharp enough to remember.

A gift that felt wrong in the hand.
A gesture that asked for gratitude but didn't contain love.
A bouquet of wilted tulips that I still thanked someone for as if it were beautiful, because the truth would have created conflict, and I had been trained to avoid conflict.

That training runs deep in women.

You learn to protect the room.

You learn to smooth the awkwardness.

You learn to make low effort feel acceptable so you don't look ungrateful.

And each time you do it, you create another kind of noise: the noise of performance.

I didn't know I was performing. I thought I was being kind.

But kindness without clarity isn't kindness. It's camouflage.

Institutional noise: the form as reality

The final kind of noise was institutional.

In my professional world, I've spent years watching organizations function on a strange combination of competence and denial. M&A, transitions, carve-outs—these are environments where "the plan" exists on paper and reality exists in the messy middle, and the distance between them is where most of the risk lives.

Institutions talk in clean sentences:

We value transparency.
We prioritize compliance.

We care about people.
We have best practices.

And then, under the surface, they optimize for different things:

liability, reputation, risk containment, short-term metrics, plausible deniability.

I learned to navigate that gap because I had to. I learned to build systems that record what is actually happening, because the official narrative often wasn't enough to protect anyone.

But even as I learned that professionally, I didn't fully apply it personally.

Because it's one thing to see institutional distortion in a company.

It's another thing to see it in your own life.

When noise becomes identity

The most dangerous thing about noise is that it doesn't just obscure reality.

It trains you to become a certain kind of person:

- the person who doesn't ask for too much
- the person who doesn't make things awkward
- the person who can tolerate misalignment and keep smiling
- the person who interprets signals generously, even when the signals repeat
- the person who keeps functioning while slowly becoming less alive

And the world rewards that person.

The world calls her "strong."
The world calls her "low maintenance."
The world calls her "so capable."

So she keeps going.

I kept going.

Until I started to notice something I couldn't unnotice:

The noise was not neutral.

The noise was instructive.

It was teaching me what I was allowed to be.

And I had accepted that curriculum without signing up for it.

THE FIRST CRACK IN THE NORMAL

The first crack wasn't a dramatic event.

It was a pattern recognition.

It was the growing clarity that certain environments—certain relationships, certain systems—required me to become smaller in order to remain "safe."

And I began to understand something that would later become the core claim of this book:

Clarity isn't comfort.

Clarity is the end of self-negotiation.

It begins when you stop calling distortion "normal."

It begins when you stop treating interruption as an accident, control as care, and performance as peace.

It begins when you notice the noise and name it.

Because once you name it, you can't unknow it.

And once you can't unknow it, a new kind of life becomes possible.

Not a louder life.

A truer one.

A life where you stop adapting to the noise and start building a ledger of what is real.

That was the beginning—before I had language for it, before I had the technology metaphors, before I could see the future clearly.

It started with the simplest, most human act:

I stopped pretending the noise was normal.

2.

Managed, Not Met

There's a particular kind of loneliness that doesn't come from being alone.

It comes from being *handled.*

From realizing you are in relationship with people who don't actually meet you—who don't step into the space with you and say, *I'm here, I see you, go on.*

Instead, they manage.

They redirect.
They regulate.
They keep you within the boundaries of what is comfortable for them.

And because "management" can look like care, it can take years to see it clearly.

THE INTERRUPTION THAT ISN'T AN INTERRUPTION

If you want to understand the difference between being met and being managed, start with something small—small enough that you can deny it for a decade.

Conversation.

A normal phone call. A normal visit. A normal exchange that seems fine until the moment you begin speaking from your center—until you become animated, alive, proud, excited, *you.*

That's when it happens.

The interruption.

Not the kind where someone is excited and overlaps you, not the kind where a thought bubbles up naturally. The other kind.

The kind that arrives like a reflex.

A comment to the dog.
A sudden change of subject.
A "oh, hold on" for something trivial.
A detour that pulls the energy away from what you were building in the air.

At first it looks accidental. Harmless. Human.

But over time it reveals itself as a pattern: it happens when the subject is you.

Not when you're talking about errands. Not when you're talking about neutral news. It happens when you are *present.*

And that's the signal:

My aliveness is not welcome here.

Or—more precisely:

My aliveness requires something from the other person that they do not want to give.

Attention.
Witnessing.
Equality.
Space.

So they regulate the moment the way people regulate discomfort: they deflect.

The cruelty of this pattern is that it is plausible. It gives no clean scene to point to.

If you name it, you risk sounding petty.

So most people don't name it.

They do what I did.

They shrink.

I would stop mid-thought, as if my own sentence had become unnecessary. I would let the detour happen. I would tell myself I didn't need to insist on being heard.

And I believed I was doing something noble.

I believed I was choosing dignity.

But the truth is that I was accommodating someone else's limitations by making myself smaller.

That's not dignity. That's adaptation.

And adaptation, repeated enough, becomes identity.

THE SOCIAL TRAINING OF WOMEN

It's almost impossible to talk about this without naming the gendered training that runs underneath it.

Women are taught to be "easy."

Easy to be around.
Easy to please.
Easy to interrupt.
Easy to redirect.
Easy to keep comfortable.

We learn early that a woman with strong presence can trigger discomfort—especially in people who unconsciously associate her presence with threat.

So we become skilled at reducing our own impact.

We learn to soften our voice.
We learn to laugh off dismissals.
We learn to act grateful for low effort.
We learn to keep the peace even when the peace is purchased with our own disappearance.

Being managed begins to feel like normal love.

And being met begins to feel suspicious, even unsafe—because it requires a kind of equality we were never trained to expect.

MONEY MEETINGS: MANAGEMENT WITH PAPERWORK

It wasn't only conversation.

There were also the money meetings.

Weekly "budget" meetings that were framed as responsible partnership. In theory, two adults sit together, look at their finances, make decisions, align.

In reality, the dynamic was never neutral.

It felt like scrutiny.

It felt like one person being evaluated and the other person being authorized.

And the strangest part—the part that shows you how little money meetings are actually about money—is that the facts didn't match the posture.

I made more salary.

I was not reckless.

I handled responsibility.

Yet the ritual remained.

And a ritual is a form of architecture. It shapes reality no matter what the words around it say.

That's what management does: it uses structure to establish hierarchy while claiming virtue.

If you are in a dynamic where the structure consistently places you on the defensive—explaining, justifying, proving—you are not being partnered.

You are being managed.

That management can be subtle. It can come with the language of care:

I'm just trying to keep us safe.
I'm just being responsible.
I'm just making sure we're aligned.

But the body knows the difference between alignment and surveillance.

Alignment feels like shared truth.
Surveillance feels like being watched.

And being watched changes a person.

Under surveillance, you either comply or you become secretive. Either way, intimacy dies.

THE GIFT THAT WASN'T A GIFT

When you've lived inside money as management, even "generosity" can become confusing.

A cash gift at Christmas, for example.

On paper: generous.

In the hand: strange.

Cash is not intimate. Cash doesn't require knowing you. It doesn't require attention to your interior world. It is a way of giving while remaining above the exchange.

And in a dynamic where money already carried hierarchy, the cash gift reads like a message:

I provide.
I decide.
I grant.

It puts you back in the position of recipient, even if you are fully capable, fully contributing, fully sovereign.

This is what management does: it manufactures dependence as a mood.

It creates an atmosphere where you feel you should be grateful for what you were already entitled to.

And if you feel grateful, you are less likely to question the structure.

THE SUBTLE SHRINK

Being managed rarely looks like a dramatic betrayal.

It looks like a slow contraction.

It looks like you choosing "not to make a big deal" so many times that eventually you can't find the part of you that wanted to speak up.

You begin to pre-edit yourself:

- Don't bring that up; it'll get awkward.
- Don't talk about your work too much; it makes them uncomfortable.
- Don't mention that you don't like the dynamic; it'll turn into a fight.

- Don't ask for more; you'll be seen as demanding.

And because the shrink is subtle, it feels like maturity.

But maturity is not self-erasure.

Maturity is truth with restraint, not truth with disappearance.

Why this matters now

I didn't understand how dangerous management was until I saw it as a pattern that extends beyond relationships.

Management is not just interpersonal.

It is cultural.

It is institutional.

It is the default mode of systems that can't tolerate a fully sovereign human being.

Systems manage people when they don't trust them. People manage others when they don't want to meet them as equals.

And meeting someone as an equal is costly.

It requires presence.

It requires humility.

It requires letting the other person exist without controlling the narrative.

Management is cheaper. It keeps power intact.

So the world offers management everywhere: in workplaces, in institutions, in families, in relationships. It becomes so common that people confuse it with love.

And that is the tragedy:

Many people never realize they are not being loved poorly.

They are being managed efficiently.

The first act of clarity

My first act of clarity wasn't confrontation.

It wasn't a big speech.

It was an internal recognition, simple but irreversible:

I can tell the difference between being met and being managed.

I can feel it in my body.

I can see it in patterns.

I can stop pretending that interruptions are random when they are timed.

I can stop calling money meetings "responsibility" when the structure is hierarchy.

I can stop performing gratitude for gestures that carry control.

That recognition doesn't fix anything immediately.

But it changes everything.

Because once you can name management, you can stop cooperating with it.

You can stop shrinking.

You can stop translating other people's limitations into your own insufficiency.

And you can begin the deeper work—more difficult than leaving any room:

You can begin building a life where you are actually met.

Not handled.

Not monitored.

Not redirected.

Met.

And that—quietly, cleanly—is where the noise begins to lose its power.

3.

The Cost of Being Reasonable

For a long time, I believed "reasonable" was a virtue.

Reasonable women don't make scenes.
Reasonable women don't demand too much.
Reasonable women interpret generously.
Reasonable women keep the peace.
Reasonable women hold the whole room together with their tone.

Reasonable women are praised.

They are trusted.
They are invited back.
They are considered safe.

And because being reasonable wins social approval, it's easy to confuse it with being free.

But being reasonable has a hidden cost.

The cost is that you start negotiating with yourself.

Not once. Not occasionally.

Constantly.

The bargain I didn't name

I didn't tell myself, *I'm going to abandon my own truth so everyone stays comfortable.*

I told myself softer things—things that sounded mature.

This isn't worth a fight.
I don't want drama.
It's not that big of a deal.
They didn't mean it.
I can handle it.
I'm fine.

Those phrases are the language of reasonableness.

They are also, often, the language of self-erasure.

I was not lying when I said them. I was doing something deeper: I was bargaining with reality.

I was buying peace in the room by spending peace inside myself.

And because the peace inside myself didn't break all at once, I didn't see the transaction.

It felt like maturity.

It was actually a slow leak.

The silence that looks like dignity

When the interruptions happened—the moment I began speaking about something that mattered to me and the energy got redirected—I didn't argue.

I didn't insist.

I didn't push my sentence back into the room.

I stopped.

I became smaller in real time.

And the story I told myself was: *I'm not going to compete for attention.*

That story contains a kind of dignity. It's true that you don't want to beg.

But there's another truth hidden beneath it:

I had learned that insisting on being heard could cost me love.

Or at least, it could cost me the illusion of harmony.

So I chose harmony.

I chose to be easy to be around.

And each time I chose that, I trained my nervous system to associate my own aliveness with relational risk.

This is the cost that doesn't show up on paper: the cost of learning to pre-edit your joy.

THE PERFORMANCE OF GRATITUDE

The wilted tulips are still a perfect symbol in my mind because they reveal the moment so clearly.

A gift that was technically a gift, but emotionally a message.

And I acted thankful.

Not because I didn't know.

Because I did know, and knowing created a problem:

If I told the truth, I would become the difficult one.
If I told the truth, I would disrupt the room.

If I told the truth, I might reveal something I wasn't ready to fully face—that the offering matched the relationship.

So I performed gratitude.

Women are trained to do this: to make the other person feel good about what they did, even if what they did didn't actually contain care.

That performance is work.

It is emotional labor.

And it trains you to accept low effort as normal while giving your own emotional sophistication away for free.

I didn't see it as labor then.

I saw it as manners.

But manners can become a mask that hides your own hunger for real love.

MONEY MEETINGS AND THE ADDICTION TO "BEING FAIR"

The money meetings taught me something else: reasonableness can become a cage made out of fairness.

Fairness sounds noble. It can be.

But in a dynamic where one person is controlling, fairness becomes a trap because it forces you into constant proof.

Proof that you're responsible.
Proof that you're not "spending too much."
Proof that you're not selfish.
Proof that you deserve autonomy.

I tried to be fair.

I tried to be transparent.

I tried to participate in the ritual as if it was just good adulthood.

And each time, I paid a cost: I reinforced the structure.

I made the dynamic look mutual when it wasn't.

And because it looked mutual, it became harder to name.

This is a specific kind of reasonableness: the kind that keeps you participating in something that harms you because you don't want to be accused of being unreasonable.

You keep showing up to court even though the court is illegitimate.

THE REAL COST: SELF-TRUST

The deepest cost of being reasonable is not the time you spend smoothing situations.

It's what happens to your relationship with your own perception.

If you override your instincts often enough, you start to distrust them.

If you keep explaining away patterns that your body is clearly registering, your internal compass begins to wobble.

You become less decisive.
Less direct.
Less willing to confront reality.

Not because you're weak.

Because you've trained yourself to prioritize social comfort over truth.

This is why reasonable people can end up in unreasonable situations for a long time.

They keep trying to solve the problem with more reasonableness.

But the problem is not a lack of reason.

The problem is misalignment.

And misalignment does not respond to politeness.

It responds to clarity.

The moment I noticed the ledger

There was a point when I began to feel a strange fatigue—not a normal tiredness, but a deeper exhaustion.

The exhaustion of constantly translating.

Translating other people's interruptions into harmlessness.
Translating control into responsibility.
Translating my own discomfort into sensitivity.
Translating the truth into something softer so no one would feel exposed.

That translation work is invisible, which is why it's so draining.

And then one day—I don't remember exactly which day, because this kind of shift isn't cinematic—I realized something that felt both obvious and shocking:

I was spending myself.

I was paying for peace with myself.

I was purchasing quiet external harmony by creating internal conflict.

I was choosing the room over my own witness.

And this is where the idea of the ledger becomes more than a metaphor.

Because a ledger is simply a record of transactions.

And I had been making transactions for years:

- trading speech for approval
- trading standards for "not being difficult"
- trading truth for temporary calm
- trading self-trust for belonging

Once I saw it as a ledger, I couldn't pretend it was just "life."

It was a system.

And systems can be changed.

PAST VS FUTURE: REASONABLENESS AS A VULNERABILITY

In the past, being reasonable was mostly a social advantage. It made you likable. It made you employable. It made you "easy."

In the future, being reasonable—without clarity—will be a vulnerability.

Because the future will be filled with systems that reward compliance while offering you narratives that sound reasonable:

This is for safety.
This is best practice.
This is policy.
This is just how things are now.

Those phrases will be easier than ever to manufacture, easier than ever to deliver with professional polish, easier than ever to back with data you can't verify.

If your instinct is always to be reasonable, you may comply with things that quietly diminish your autonomy.

Not because you agree with them.

Because you've been trained not to make trouble.

So clarity becomes the upgrade.

Clarity isn't loud rebellion.

Clarity is simply refusing the transaction:

I will not buy peace by selling pieces of myself.

The first refusal

The shift didn't happen through confrontation.

It happened through refusal.

I began refusing internally first—refusing to rewrite what I could see.

When someone interrupted, I stopped calling it nothing.
When something felt controlling, I stopped calling it care.
When my body tightened, I stopped calling it sensitivity.
When a gesture felt like positioning, I stopped calling it generosity.

That refusal is a kind of quiet bravery.

Because it ends the fantasy that you can stay comfortable and also stay true.

Sometimes you can.

Often, you can't.

But here's what I learned:

When you stop paying for peace with yourself, you may lose a certain kind of harmony.

And what you gain is something more valuable:

You gain coherence.

You gain the right to live inside your own reality.

You gain the calm that comes from not negotiating your own witness.

That calm is not dramatic.

It is steady.

It is clean.

And it is the beginning of a different kind of life—one where being "reasonable" is no longer the goal.

Being true is.

The Self-Negotiation Loop

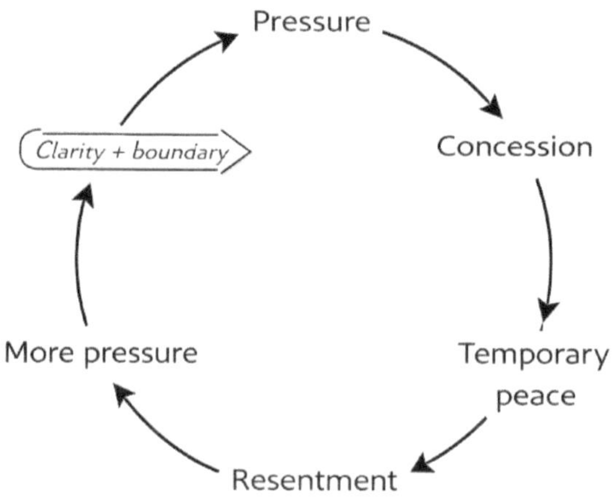

Tech Interlude A

The First Edit

The first time reality gets edited, it doesn't look like censorship.

It looks like convenience.

It looks like credentialing.
It looks like a login screen.
It looks like an "official" record.
It looks like a feed that decides what matters.
It looks like a form that turns a human life into checkboxes.

It looks like modern life.

We don't notice the edit because we inherit it. We're born into a world where truth is increasingly mediated by systems—systems that don't witness, don't love, don't remember like humans do. They recognize. They validate. They approve. They deny.

And slowly, without anyone announcing it, a new kind of reality settles over us:

Reality becomes what the system recognizes.

Not because the system is always right, but because the system controls access—to money, to opportunity, to reputation, to legitimacy.

So we adapt.

We learn to speak in the system's language.

CREDENTIAL TRUTH

The modern mind is trained to outsource reality to credentials.

If a person has the degree, they are "qualified."
If they have the title, they are "credible."
If they have the institution behind them, they are "real."

Credentials aren't inherently bad. Sometimes they represent rigor. Sometimes they protect standards.

But they also do something subtle to the psyche: they make us treat recognition as truth.

We begin to doubt ourselves until the system confirms us.

We defer to people because their letters and logos are louder than our own experience.

We keep trying to become legible—approved—because approval feels like safety.

And that's the first edit: the internal edit.

The edit that says: *My perception is not enough.*

Feed truth

Then came feeds.

The stream of curated reality that never ends.

Feeds don't just show you information. They shape what you think exists.

They teach you what is worthy of attention and what can be ignored.

They train you, quietly, to feel real only when mirrored back—through likes, comments, views, engagement.

And because feeds operate at the level of emotion, they become a truth machine without calling themselves one.

If it is trending, it is "real."
If it is viral, it is "important."
If it is unseen, it is "nothing."

A feed is not reality, but it trains the nervous system to treat it like reality.

That's the second edit: the attention edit.

The edit that says: *If it isn't surfaced, it doesn't matter.*

Institutional truth

Institutions complete the training.

They don't just manage resources. They manage legitimacy.

They create records. They issue IDs. They certify eligibility. They define what counts.

They turn human complexity into administrative reality:

- The form says you are married or not.
- The system says you are approved or denied.
- The record says you are employed or not.

- The database says you are compliant or not.

Institutions are built for scale. Scale requires categories. Categories require simplification.

So the institution doesn't meet you—it processes you.

And if you don't fit, the system doesn't necessarily argue with you. It simply fails to recognize you.

You become invisible in a way that feels personal even when it isn't.

That's the third edit: the identity edit.

The edit that says: *You are what we can classify.*

The new religion: "If the system doesn't confirm it, it isn't real."

This is the modern replacement for older forms of meaning.

Where past generations might have anchored truth in family, community, faith, direct witness—

we anchor it in:

- official documentation
- platform legitimacy
- institutional consensus
- algorithmic visibility

It feels rational because it's external and standardized.

But it creates a quiet fragility:

When the system is wrong, people doubt themselves.
When the system is silent, people disappear.
When the system changes, people lose their footing.

And because most systems are opaque, the user can't audit the decision. They can only experience the outcome.

Approved.
Denied.
Shadowed.
Deprioritized.
Flagged.
Frozen.

No human explanation. Just a status.

Why this matters now

For a while, we could live with these edits because the world still had enough friction.

A human could still appeal to a human.
A reputation still lived partly in real rooms.
A story still required effort to manufacture.

But friction is dropping.

AI makes language cheap and perfect.
Automation makes decisions fast and invisible.
Data makes people legible in ways they don't control.

This means the system's edits will become more frequent—and harder to detect.

The first edit was accepting that recognition equals truth.

The next edit is believing the output without seeing the mechanism.

That is where people begin to lose their internal witness.

A private ledger

The antidote isn't paranoia.

It's not rejecting all systems.

The antidote is retaining something ancient and necessary:

A private ledger.

A record of what you know.
A discipline of remembering accurately.
A willingness to trust your own perception before the system blesses it.

Because if reality becomes only what the system recognizes, then you will spend your life trying to become legible.

And legibility is not the same as truth.

The ledger is how you keep truth alive.

It is how you resist the first edit.

Not by shouting.

By witnessing.

By refusing to let the system's mirror replace your own eyes.

ACT II

THE MIRROR BREAKS

4.

When the Forms Stopped Coming

The old contract was never written down.

But I believed in it as if it were law.

If I do the right things, I'll be safe.
If I work hard, I'll be protected.
If I'm reasonable, I'll be treated fairly.
If I follow the process, the process will hold.

Most adults live by some version of this contract. It's what makes a complicated world feel navigable. It's what allows you to wake up and build a life without being crushed by the infinite ways that life can break.

And for a long time, the contract seemed to work—just enough to keep my faith intact.

Until the day it didn't.

Not with a dramatic betrayal. Not with a singular catastrophic event.

With something quieter.

The forms stopped coming.

The System Goes Silent

It's hard to explain to someone who hasn't felt it how unnerving institutional silence is.

When a person goes quiet, you can interpret it a hundred ways: they're busy, they're upset, they're distracted, they're avoidant, they're afraid.

But when a system goes quiet—when the expected communication doesn't arrive, when the confirmation doesn't appear, when the next step doesn't materialize—it feels like reality itself has paused.

Because forms are not just paper.

Forms are acknowledgment.

Forms are the system saying: *We see you. We recognize you. You exist inside our record.*

So when the forms stop coming, a specific kind of modern panic appears:

What am I if I'm not being processed?
What happens if no one confirms I'm real?
Where do I go if the institution doesn't respond?

It's not logical panic. It's bodily.

The nervous system has been trained to equate "recognized" with "safe."

And once you see that, you realize how deep the training goes.

It wasn't one form. It was the pattern of permission.

The first time, you tell yourself it's a glitch.

The mail was slow. The admin is behind. The office is busy. The person on the other end is out sick.

You do what "reasonable" people do: you wait.

Then you follow up.

Then you wait again.

Then you start gathering your documentation, as if you must prove you deserve to be answered.

Then you soften your tone—because you don't want to irritate the gatekeeper.

Then you become polite enough that you are nearly invisible.

This is what modern institutions do to people: they train them to self-diminish in the pursuit of recognition.

And the hardest part is that no one is "doing" it to you in a way you can point to.

It's just how the machine works.

The machine doesn't hate you.

The machine simply doesn't know you.

And because it doesn't know you, it cannot witness you.

It can only process inputs.

So when the input doesn't fit, it delays.

It loops.

It rejects.

It goes silent.

THE SPIRITUAL EFFECT OF BUREAUCRACY

When I say "the forms stopped coming," I'm not only describing paperwork.

I'm describing what happens when the world you trusted no longer provides a coherent reflection.

A missing form becomes a missing mirror.

And suddenly you understand something you didn't want to understand:

Most people don't know who they are without institutional confirmation.

We want the system to validate our pain.
We want the system to validate our success.
We want the system to validate our worthiness.
We want the system to validate our identity.

We are modern humans, and we have been trained to treat the institution's output as truth.

Approved. Denied. Eligible. Ineligible. Verified. Flagged.

That language doesn't just regulate transactions.

It regulates the self.

So when the forms stop coming, it doesn't feel like a delay.

It feels like erasure.

And that is why people tolerate so much: because the fear beneath many adult compromises is not failure.

It is disappearance.

I RECOGNIZED THE SAME PATTERN IN MY OWN LIFE

The more I watched institutions, the more I realized I'd been living a version of the same dynamic in personal relationships.

When you are with someone who manages you instead of meeting you, they also control "forms."

They decide what gets acknowledged and what gets ignored.

They decide what counts as real.

They may not deny you openly. They just interrupt. Redirect. Go quiet. Change the subject. Treat your inner world as optional.

The human version of "forms stopped coming" is being in a room where your reality isn't received.

Where your joy isn't held.

Where your truth becomes inconvenient.

And because you don't want conflict, you adapt.

You become small. You become "reasonable." You stop expecting the form to arrive.

At some point, you stop mailing the request.

That is how people vanish inside their own lives.

The institution does it with silence.

The relationship does it with interruption.

The result feels the same: you become less legible, even to yourself.

THE COLLAPSE OF THE OLD CONTRACT

What broke in me wasn't just trust in the system.

It was trust in the contract.

Because once the forms stop coming, you realize:

Doing the right things does not guarantee safety.
Following the process does not guarantee fairness.
Being reasonable does not guarantee being seen.

That realization is painful.

It feels like losing a religion.

It can also feel like liberation.

Because the old contract has a hidden cost: it makes you obedient to outcomes you don't control.

It trains you to keep behaving "correctly" even when the environment is wrong.

And when you finally see the contract for what it is—a coping mechanism—you stop spending your life trying to earn safety through compliance.

THE PIVOT: FROM PERMISSION TO PROOF

This is where the ledger begins.

Not yet as Bitcoin or technology, not yet as ideology.

As a human need.

If the system can go silent, I need a record that doesn't require the system.

If the institution can fail to recognize me, I need a form of recognition that I can't be denied.

If the world can edit reality, I need a way to keep reality intact.

This is the pivot from permission to proof.

Permission is when your reality depends on someone else's gate.
Proof is when your reality can stand without anyone's blessing.

I didn't immediately know how to build proof.

At first, I only knew how to notice the silence.

But noticing is not nothing.

Noticing is the beginning of sovereignty.

Because the moment you realize the forms are not guaranteed, you stop living as if you are safe by default.

You stop waiting for the system to make sense.

You start making sense on your own.

The first honest sentence

The first honest sentence I said to myself was simple:

The system doesn't have to be fair to be powerful.

Once you accept that, you stop expecting fairness as evidence of truth.

You stop interpreting silence as your failure.

You stop begging.

And you begin doing the most radical thing a modern adult can do:

You begin to become your own witness.

Not through anger.

Through record.

Through clarity.

Through the decision to keep reality intact even when no one sends the form.

That is what broke.

And that is what began.

5.

The Day I Stopped Explaining

There was a day—more like a threshold—when I realized I had become fluent in translation.

Not language translation.

Reality translation.

I could take something sharp and make it soft enough to swallow.

I could take something humiliating and rename it "not a big deal."
I could take something controlling and call it "responsibility."
I could take someone's absence and call it "they're busy."
I could take my own discomfort and call it "sensitivity."

I was so good at this that I didn't recognize it as a skill.

I recognized it as maturity.

But maturity isn't the ability to endure distortion.

Maturity is the ability to look directly at what is happening without needing to protect the story.

The day I stopped explaining was the day I stopped protecting the story.

EXPLAINING IS A KIND OF ANESTHESIA

Explaining is not always wrong. Sometimes it's compassion. Sometimes it's context. Sometimes it's the intelligence of not being reactive.

But there's a version of explaining that functions like a sedative.

It doesn't clarify. It numbs.

It's the kind of explaining that isn't meant to help you understand someone—it's meant to help you avoid what you already understand.

Because what you already understand would require change.

So you explain instead.

You smooth the edges.
You soften the impact.
You keep the narrative intact.

And you keep living inside a life that doesn't fit.

I didn't wake up one morning and decide to anesthetize myself.

I learned it.

I learned it in relationships, where seeing too clearly would have made me "difficult."
I learned it in institutions, where naming the gap between policy and reality would have made me "uncooperative."
I learned it in culture, where women are rewarded for being calm, agreeable, reasonable—even when they are swallowing their own truth.

Explaining kept me socially safe.

It also kept me internally split.

THE MOMENT YOU REALIZE YOU ARE REWRITING YOUR OWN LIFE

There's a particular feeling that comes right before a major shift.

It's not dramatic.

It's a quiet nausea.

A sense that your mind is trying to tell you something your mouth doesn't want to say.

I felt it most often after conversations where I had disappeared.

A call with my mother, for example, where everything is normal until I begin talking from my center—something I'm building, something that matters, something that is mine—and then the interruption arrives: talking to the dog, pivoting the subject, breaking the thread.

And I would do what I had always done.

I would stop.

I would become smaller.

Then later, I would explain it away:

She didn't mean it.
She's distracted.
It's nothing.
It's just how she is.

What I didn't say—what I couldn't admit for a long time—was the simplest truth:

When I become fully myself, she leaves.

Not physically.

Energetically.

Presence withdraws.

Witnessing disappears.

The system goes silent.

And the "explaining" wasn't helping me love her better.

It was helping me avoid what that pattern meant.

Because what it meant was painful:

It meant I could not get certain needs met there.

It meant I would have to stop trying.

It meant I would have to grieve the fantasy that if I said it "just right," I would finally be received.

Explaining protected the fantasy.

Clarity ends fantasies.

The money story I kept smoothing

Explaining also showed up in the places where power hides best: money.

Weekly budget meetings can be normal. They can be healthy.

But when a financial ritual feels one-sided—when it consistently places you in a defensive posture, justifying, accounting, proving—you are not participating in partnership.

You are being monitored.

And I explained that away too.

This is responsibility.
This is what adults do.
This is what good couples do.
It's for safety.

Even when my body knew the truth: it was hierarchy disguised as virtue.

The same goes for gifts that were technically generous but emotionally strange—cash as a present inside a dynamic where money already carried control.

Explaining turns strangeness into normal.

Explaining turns control into care.

Explaining lets you remain "reasonable."

But reasonableness, repeated long enough, becomes complicity with your own diminishment.

The real reason people explain

Here's what I learned, slowly, without glamorizing it:

People don't explain because they're naïve.

They explain because the alternative is grief.

If you stop explaining, you may have to admit:

- this relationship cannot hold me
- this system is not designed to be fair
- this person is not coming back emotionally
- this "gift" is not love
- this dynamic is not going to change because I behave better

Explaining is a way of postponing that admission.

And postponing grief can feel like survival.

But the debt accumulates.

The Day the Pattern Became Undeniable

I can't point to one perfect scene and say, *this is when it happened.*

But I can describe the internal moment.

I remember catching myself mid-explanation—mid-translation—and feeling something in my chest tighten, as if my body was saying:

Stop lying to protect them.

Not "lying" in the moral sense.

Lying in the spiritual sense—the act of bending reality to preserve comfort.

I realized that I had been working so hard to be fair to everyone else's motives that I had become unfair to my own experience.

I gave endless benefit of the doubt outward.

I offered almost none inward.

And I finally saw the cost:

I was losing my internal witness.

I was becoming less clear, less decisive, less alive—not because I lacked intelligence, but because I was constantly editing what I knew.

So I did something small that felt enormous.

I stopped.

I stopped interpreting patterns as accidents.

I stopped smoothing the sharp edges of reality.

I stopped telling myself the discomfort was nothing.

I stopped rewriting my own experience so the room could remain calm.

WHAT REPLACES EXPLANATION ISN'T CRUELTY—IT'S RECORD

People worry that if they stop explaining, they'll become cold.

They won't.

What replaces compulsive explaining isn't harshness.

It's accuracy.

It's the ability to say:

This happened.
This repeats.
This affects me.
This costs me.

That's the beginning of the ledger.

Because explanation is fluid. It changes with mood. It changes with hope. It changes with fear.

Record is stable.

Record is how you remain coherent.

Record is how you stop being gaslit—by others, yes, but more importantly by yourself.

PAST VS FUTURE: EXPLANATION BECOMES MORE DANGEROUS

In the past, compulsive explaining mainly hurt you privately.

It kept you stuck. It delayed decisions. It softened boundaries.

In the future, compulsive explaining will become easier to exploit.

Because the world is moving toward systems that can generate plausible stories at scale:

- polished language
- soothing narratives

- "best practice" rationales
- professional certainty
- algorithmic explanations that sound neutral

You will be offered infinite explanations for why something is necessary, why something is restricted, why something is delayed, why you were deprioritized, why you should comply.

The more persuasive the world becomes, the more valuable your internal witness becomes.

Clarity will belong to the people who can hear an explanation and still ask:

What is the pattern?
What is the incentive?
What is the cost?
What is actually true?

THE FIRST ACT OF SOVEREIGNTY

The day I stopped explaining, I didn't become louder.

I became cleaner.

I didn't confront everyone.

I stopped negotiating internally.

I began to live as if my own perception mattered.

And that shift changed everything.

Because once you stop rewriting your experience, you stop living inside other people's reality.

You stop waiting for the form.

You stop begging for recognition.

You become your own witness.

And when you become your own witness, something quiet but irreversible happens:

You can no longer be managed by narratives that require your self-erasure.

You may still love people.

You may still have compassion.

You may still choose restraint.

But you will stop editing the truth to preserve comfort.

That is not cruelty.

That is the beginning of a life that belongs to you.

6.

The Ledger Appears

The ledger didn't appear as an idea.

It appeared as a need.

A need so basic that once I felt it, I couldn't unfeel it: I needed a way to keep reality intact.

Because once you stop explaining—once you stop rewriting what you know to make the room more comfortable—you notice how slippery modern life really is.

Words change.
Stories change.
People revise their own past.
Institutions update their policies as if the update erases what happened yesterday.

The same behavior is "fine" one day and "unacceptable" the next.
The same truth is celebrated in one room and punished in another.

And somewhere inside all that drift, the self can begin to wobble.

So the ledger appears as a stabilizer.

Not a spreadsheet. Not a weapon.

A record.

Before it was explicit, it was already there

This is important: the ledger wasn't new. I had been keeping it my whole life.

Most people are.

Your body keeps a ledger whether you want it to or not.

It records the moment you were interrupted when you were shining.
It records the moment you swallowed your truth to keep peace.
It records the tension in a conversation where you were being evaluated.
It records the subtle humiliation of being managed under the language of care.
It records the strange emptiness after you perform gratitude for something wilted.

The problem isn't that the ledger doesn't exist.

The problem is that we're trained to distrust it.

We're trained to call it "overthinking."
We're trained to call it "being sensitive."
We're trained to call it "making a big deal."

So we override it.

We teach ourselves to ignore the entries.

But after a certain point, ignoring the entries becomes unbearable.

Because the entries don't stop arriving.

And the cost of ignoring them is coherence.

THE MOMENT I STARTED WRITING IT DOWN

There is a quiet kind of power in writing something down.

Not for social media. Not for anyone else.

Just for truth.

Writing down what happened forces you to treat your own perception as real.

It's the opposite of gaslighting.

It is the simplest form of self-respect.

I began to record patterns.

Not with dramatic language—just clean notes.

- *She interrupts when I talk about what excites me.*
- *That meeting wasn't mutual; I left feeling monitored.*
- *That gift felt like positioning.*
- *My body tightened before I said yes.*
- *The system went silent after I asked for clarity.*

At first it felt almost childish, like keeping receipts in a drawer.

Then I realized: that's exactly what it is.

Receipts aren't bitterness.

Receipts are protection against revision.

Because people revise.

They revise their motives. They revise the past. They revise what they said. They revise what they promised. They revise what happened if revising makes them look better.

Sometimes they revise consciously.

Often they revise because their nervous system can't tolerate accountability.

Institutions do the same thing—just with better branding.

They call it a "policy update."

The Ledger Becomes an Internal Architecture

When you make your internal record explicit, something changes.

You stop living in a mood-based reality.

You stop letting the last conversation overwrite the whole pattern.

You stop letting a single nice moment erase the repeated signal.

The ledger holds the full truth at once:

Yes, there were moments of warmth.
And yes, the interruptions repeat.
Yes, the person can be kind.
And yes, the dynamic shrinks me.
Yes, the institution has good intentions.
And yes, it optimizes for itself.

The ledger doesn't demonize.

It integrates.

This is what clarity actually is: the capacity to hold multiple truths without collapsing into denial.

The first thing the ledger gave me: self-trust

I didn't realize how much self-trust I had lost until I started getting it back.

Self-trust isn't bravado. It isn't "confidence."

It's the simple ability to believe your own experience.

When I made the ledger explicit, I stopped asking myself to forget what my body already knew.

I stopped bargaining.

I stopped negotiating.

I stopped trying to *prove* that my discomfort was legitimate.

I began to treat my internal signals as data.

Not data in a cold way.

Data in a merciful way.

Because mercy is not pretending nothing happened.

Mercy is being honest enough to stop repeating harm.

Why the ledger matters in love

Here's the part people misunderstand: the ledger doesn't make you less loving.

It makes your love more real.

Without a ledger, love becomes a fantasy project.

You love what you hope someone will become.
You love what you wish they could give.
You love the version of the relationship that exists only when you are editing yourself.

With a ledger, love becomes an act of consent.

You can love someone and still see them clearly.

You can forgive and still remember.

You can understand someone's limitations and stop handing them access to your deepest parts.

Because your ledger is not an argument.

It's a boundary made out of truth.

WHY THE LEDGER MATTERS IN MONEY

It was around this time that the metaphor started widening.

Money is a ledger. We all know that, intellectually.

But I began to feel it viscerally: the difference between a ledger that can be edited and one that can't.

A marriage can become an editable ledger: one person can rewrite what was agreed, what was fair, what was "too much," what was "responsible."

An institution is an editable ledger: rules change, definitions shift, decisions get buried inside "process."

A reputation is an editable ledger: a narrative can be altered overnight by what's amplified, what's hidden, what's repeated.

This is why the ledger principle isn't just personal.

It's structural.

You can't build a coherent life on ledgers that are constantly rewritten by others.

At some point, you need an anchor.

You need a record that doesn't depend on mood, status, or permission.

That is why the idea of verifiable ledgers—hard, auditable records—hits something deep in me.

Not as politics.

As psychology.

As sanity.

PAST VS FUTURE: THE LEDGER BECOMES SURVIVAL

In the past, you could get away with a softer relationship to truth.

Reality had friction.

There were fewer channels. Fewer narratives. Fewer ways to fabricate the appearance of certainty.

People could still meet in rooms and feel what was real.

In the future, the world becomes smoother and more synthetic.

Words become perfect.
Images become believable.
"Evidence" becomes manufacturable.
Consensus becomes easier to engineer.
Institutions become more automated and opaque.

In that environment, an internal ledger becomes survival.

Not because you're trying to win an argument.

Because you're trying to keep your mind.

The future will reward people who can be persuaded.

It will protect people who can verify.

Verification isn't only technical. It's also personal:

- *What do I know from direct experience?*
- *What is the pattern over time?*
- *What does my body register before my mind explains it away?*

- *What did I write down when I was clear?*

That's the ledger.

The Ledger's Quiet Promise

When the ledger appeared, I stopped asking the world to hold my truth for me.

I stopped expecting relationships to validate what they could not witness.

I stopped expecting institutions to be fair simply because they were official.

I began to trust the one witness I had been trained to doubt:

My own.

And something softened in me—not into naivete, but into steadiness.

Because the ledger does not need to be defended.

It doesn't argue.

It simply holds.

It holds the record of what happened.

It holds the costs.

It holds the patterns.

It holds the truth long enough for you to make a decision that isn't based on hope or fear, but on reality.

That's what the ledger is.

Not a punishment.

A protection.

An internal architecture that says:

You will not disappear here.

And once you have that, the old contract collapses completely.

Because you no longer need the form to arrive to know what is true.

You have your own record now.

You have your own witness.

The ledger has appeared.

Editable World	Ledger World
Narratives shift	Constraints
Incentives distort	Proof
Memory is social	Accountability
	Time-stamped truth

Clarity is the bridge.

Tech Interlude B

From Paper to Data

At some point, the forms didn't stop coming.

They just stopped arriving where you could see them.

They moved behind the walls.

The modern world still runs on forms—still runs on permission, eligibility, classification, approval—but the paper version was only the training wheel. Paper was slow. Paper was visible. Paper created friction. Paper gave you a person to call.

Data is different.

Data is quiet.

Data is instantaneous and everywhere.
It is collected without asking.
It is shared without being felt.

It is scored without being explained.
It is acted on without a conversation.

And once life becomes datafied, the form becomes invisible.

The form is no longer something you fill out.

The form is you.

THE NEW FORM: A RECORD THAT FOLLOWS YOU

In the paper era, your "record" lived in folders.

A file cabinet. A manila envelope. A record you could request, sometimes even hold in your hands.

In the data era, your record lives in systems you will never see:

- databases
- identity graphs
- device fingerprints
- payment rails
- risk models
- customer profiles
- compliance systems
- recommendation engines

Your modern life generates a continuous form in real time.

What you click.
What you buy.
Where you travel.
Who you message.
How often you change phones.
Which addresses you use.

How quickly you pay.
How often you move money.
What you search.
What you linger on.

This isn't speculation. This is the business model of the digital economy.

The difference is not whether it happens.

The difference is whether you recognize it as a form.

Because if you don't recognize it, you keep thinking you are navigating life.

In reality, life is being classified as you navigate it.

FROM HUMAN JUDGMENT TO AUTOMATED JUDGMENT

In older systems, even unfair ones, there was often a human moment.

A person read your application.
A person reviewed your file.
A person decided your eligibility.
A person might be biased, but a person could also be appealed to.

Automation removed the person and replaced them with a decision pipeline.

This pipeline is not "evil."

It is optimized.

And optimization is its own ideology.

Automation optimizes for:

- speed
- consistency
- risk reduction
- cost reduction

- liability management
- policy enforcement

Notice what's not on that list:

- your nuance
- your story
- your intent
- your transformation
- your context
- your humanity

So the new system doesn't "judge" you the way a person judges.

It filters you.

It routes you.

It flags you.

It categorizes you into paths.

Approved.
Denied.
Review.
Hold.
Verify.
Wait.
Deprioritize.

And because the decision is produced by a model, the model is treated like objectivity—even when the model is simply encoding past decisions, past patterns, past institutional risk tolerance.

This is how data becomes destiny.

Not because the model is always right.

Because the model becomes the gate.

The New Gatekeeper: The Algorithm

When the gatekeeper is a person, you can speak.

When the gatekeeper is an algorithm, you can only comply.

You can only submit another document.

You can only click "appeal" and wait for a response generated by another decision layer.

This is the modern version of "the forms stopped coming."

It's not that nothing happens.

It's that what happens is happening somewhere you cannot enter.

The gatekeeper has no face.

So your relationship to power becomes abstract, and abstraction makes people feel helpless.

That helplessness is not accidental.

It is one of the emotional side effects of invisible governance.

Scoring: The Quiet Replacement for Conversation

Underneath the algorithmic gate is a deeper shift: scoring.

Scoring is the system's way of making human complexity legible.

Credit scoring is the obvious example, but scoring now extends far beyond credit:

- fraud scoring
- AML / compliance risk scoring

- account risk scoring
- transaction risk scoring
- identity confidence scoring
- content trust scoring
- reputation scoring
- moderation scoring
- recommendation scoring

You don't need to believe in an official "social credit" system to understand what this is.

It's already here in fragments.

And fragments are enough, because fragments still affect access.

Access to:

- accounts
- payment platforms
- loans
- travel
- jobs
- visibility
- distribution
- reputation

Scoring replaces a human relationship with a system relationship.

A person used to say, "Tell me what happened."

Now the system says, "Your score triggered a hold."

That's not an explanation.

That's a verdict.

Why the new form feels like gaslighting

In the paper era, if you were denied, you could often see the reason.

A missing document. A checkbox. A signature. A stated policy.

In the data era, you may be denied because of a pattern you can't see:

- your behavior looks statistically similar to fraud
- your transaction resembles a risky cluster
- your identity signals don't match expected confidence
- your activity triggered a threshold
- the platform's risk tolerance shifted this quarter
- the model updated

So you ask: *Why?*

And you get a response that feels like a polite void:

For your security…
Due to policy…
We can't disclose…
This decision is final…

This is why people start to feel insane.

Not because they are unstable.

Because they are living inside a reality that can act on them without being accountable to them.

That is the psychological effect of opaque systems:

They can change your life without having to explain your life.

The human cost: living in a world that can't meet you

This is where the theme of the book deepens.

Being managed instead of met is not just a personal relationship problem.

It is the architecture of modern systems.

Systems manage risk; they don't meet people.

They manage exposure; they don't witness truth.

They manage liability; they don't hold nuance.

Which means the world is becoming more "managed" at every layer:

- in relationships, via subtle control and interruption
- in institutions, via policy and process
- in technology, via scoring and automation

So the question becomes:

What does a human being need to stay coherent inside that world?

And the answer is the same answer you arrived at personally:

A ledger.

A record.

A way of keeping reality intact when external systems are opaque and editable.

The response is not paranoia. It's literacy.

The goal isn't to fear technology.

The goal is to become literate in what changed.

Because once you understand that the form became data and the gatekeeper became an algorithm, you stop taking the outputs personally.

You stop begging the system to "see you."

You stop expecting the machine to behave like a witness.

And instead, you begin building a life that protects what only humans can protect:

- self-trust
- relational truth
- consent
- dignity
- coherence under pressure

The old world trained us to fill out forms.

The new world trains us to become forms.

Clarity begins the moment you realize that.

And sovereignty begins when you decide you will not let an invisible record become the only story of who you are.

ACT III

THE TECHNOLOGY TURN

7.

The Age of Synthetic Signal

I used to think the most dangerous lies were the obvious ones.

The ones with contradictions you could point to.
The ones delivered by people who were clearly manipulative.
The ones that made your stomach drop immediately.

But the next era isn't built on obvious lies.

It's built on *plausible ones.*

The future belongs to synthetic signal.

Signal that looks like truth.
Feels like sincerity.
Sounds like intelligence.
Has all the right tone.
Uses all the right language.

And costs almost nothing to produce.

THE WORLD USED TO HAVE FRICTION

For most of human history, words were expensive.

Not monetarily—socially.

To craft a persuasive message, you needed time, literacy, talent, rehearsal, access. To appear coherent in public, you had to cultivate coherence in private. To sustain a lie, you had to maintain it with effort.

Words had friction.

Even manipulative people couldn't generate infinite versions of themselves.

They had to choose.

They had to commit.

And because they had to commit, patterns eventually emerged. You could read them. You could sense the gaps.

Then the friction started falling.

First with the internet—suddenly anyone could publish.

Then with social media—suddenly anyone could perform.

And now with AI—suddenly anyone can manufacture *credible language* at scale.

That's the shift.

Not that people will "lie more."

That the cost of producing persuasive signal has collapsed.

SYNTHETIC SIGNAL ISN'T JUST FAKE—IT'S OPTIMIZED

"Fake" implies crude imitation.

Synthetic signal is not crude.

It is optimized.

AI can produce language that sounds:

- thoughtful
- emotionally intelligent
- balanced
- compassionate
- confident
- authoritative
- calm
- humble
- visionary

It can generate apologies that feel sincere.

It can generate mission statements that sound noble.

It can generate explanations that sound reasonable.

It can generate "clarity" in the form of words—without the internal coherence that makes clarity real.

This is the part that matters:

AI can generate the *appearance* of clarity.

And when the appearance becomes abundant, the real thing becomes rare.

THE FIRST TIME YOU FEEL IT, IT'S DISORIENTING

There's a moment that happens to everyone now.

You read a message and think: *That's well said.*

You feel moved.

You think the person is mature, reflective, emotionally evolved.

And then you realize they didn't write it.

Or you realize they wrote it with help.

Or you realize the "person" might not be a person at all.

That moment is more than a tech moment.

It's a trust moment.

It's the moment the mind realizes:

I can no longer use words as evidence of character.

Words used to be a decent proxy for inner life.

Now they are not.

Now they can be rented.

That realization breaks something subtle in society.

Because society has always depended on proxies.

PERSUASION BECOMES CHEAP, AND THAT CHANGES EVERYTHING

When persuasion is cheap, the whole economy of trust changes.

In the old world, a polished message signaled effort.

In the new world, a polished message might signal nothing at all.

It might signal:

- a template
- a model
- a prompt

- a content pipeline
- a brand machine
- an automated persona

And because the output can be infinite, the environment becomes saturated.

You are flooded with:

- compelling arguments
- convincing narratives
- emotionally tailored stories
- authoritative confidence
- "reasonable" explanations
- perfectly calibrated tone

This is where people start to feel numb.

Not because they lack intelligence.

Because the mind cannot metabolize infinite persuasion.

So it shuts down.

It becomes cynical.

Or it becomes pliable.

Those are the two common outcomes of living inside a persuasion-rich environment:

1. **Cynicism:** nothing feels real, so nothing matters.
2. **Pliability:** everything feels real, so anything can matter.

Neither is clarity.

I RECOGNIZED THE PATTERN BEFORE I HAD THE LANGUAGE

The reason I can name this era is because I lived its precursor in human form.

I lived inside relationships and systems where words were used as management.

Not always maliciously.

But strategically.

Words that sounded responsible, while the structure was controlling.

Words that sounded caring, while the dynamic was hierarchy.

Words that sounded fair, while the reality was one-sided.

I learned to stop trusting tone.

I learned to trust cost.

I learned to watch what a person would do when no one was clapping.

Those lessons weren't just personal—they were training for this era.

Because AI is turning the whole world into the relationship dynamic I learned to leave:

a world where words can be beautiful while reality is misaligned.

THE DEATH OF "VIBES" AS TRUTH

This is going to be painful for a while: we've been trained to use vibes as a truth detector.

How a message makes you feel.
How aligned the language sounds.
How confident the speaker seems.
How soothing the tone is.

But in the age of synthetic signal, vibes are easy to engineer.

Not only by bad actors.

By anyone who wants to win.

A company can manufacture trust at scale.

A politician can manufacture sincerity at scale.

An influencer can manufacture intimacy at scale.

A stranger can manufacture emotional safety at scale.

And the human nervous system—wired for face-to-face cues—can be hijacked by language alone.

This is why the age of synthetic signal isn't mainly a tech problem.

It's a human perception problem.

What remains real when words become cheap

When words become cheap, what remains real is what cannot be generated without cost.

Consistency over time.
Sacrifice.
Accountability.
Presence.
Skin in the game.
The willingness to hold truth when it's inconvenient.

In other words:

Proof of work.

And clarity is the internal form of proof of work.

Because clarity isn't a sentence.

Clarity is the coherence that exists before the sentence.

Clarity is what prevents you from saying one thing and doing another.

Clarity is what makes you recognizable to yourself.

AI can generate a beautiful explanation.

It cannot generate your integrity.

PAST VS FUTURE: FROM SCARCITY OF SPEECH TO SCARCITY OF COHERENCE

In the past, speech was scarce enough that it could still function as signal.

If someone wrote well, it often meant something.

If someone apologized well, it often meant something.

If someone sounded wise, it often meant something.

In the future, those outputs will be abundant.

And the scarcity will shift.

The scarcity will be:

- a coherent human being
- a verifiable record
- a pattern of truth that persists across pressure
- a life whose ledger matches its language

This is the turn.

This is the moment where clarity moves from "personal growth" to "strategic asset."

Because in an editable world, the person who can't be edited—internally—has an advantage that cannot be faked.

THE FIRST RULE OF THE NEW ERA

Here is the first rule I learned, and it is the rule that will protect you:

Stop using language as proof.

Use it as information.

But don't let it be proof.

Proof is what costs something.

Proof is what endures.

Proof is what holds when no one is watching.

In the age of synthetic signal, persuasion will be cheap.

So your attention becomes expensive.

Your trust becomes expensive.

Your clarity becomes priceless.

And the story of this book—my story, really—is the story of learning to protect that clarity long before the world made it necessary.

Because once you see the synthetic signal for what it is, you stop being hypnotized by perfect words.

You start listening for something quieter.

The sound of reality.

The sound of cost.

The sound of a ledger that matches the life.

That sound is rare now.

It will be rarer soon.

And it will be the difference between being managed by narratives and living inside what is true.

8.

The End of Trust-by-Default

There was a time when trust was the default setting.

Not because people were better.

Because reality had limits.

A photo meant something because it was hard to fake.
A letter meant something because someone had to write it.
A recommendation meant something because it traveled through real relationships.
A resume meant something because lying required nerve and risk.
A public statement meant something because it could be challenged in a shared world.

Even consensus—imperfect, often unfair—had a kind of gravitational force because the inputs were constrained.

Reality had friction. Evidence had weight.

And then the constraints started dissolving.

Now we are entering a world where the outputs are no longer proof of anything.

Not because all outputs are fake.

Because any output *could be.*

That is the subtle but profound shift.

When everything can be generated, trust-by-default collapses.

TRUST USED TO BE A SOCIAL LUBRICANT

Trust-by-default is how society scales.

You don't verify every receipt.
You don't audit every interaction.
You don't confirm every identity.

You assume—within reason—that the world is mostly real.

That assumption is not naïveté. It's a survival mechanism.

If you had to verify everything, you would never leave your house.

So we evolved social shortcuts:

- credentials
- brands
- institutions
- norms
- reputations

- "common knowledge"
- consensus

These shortcuts were never perfect, but they made life possible.

They were the shared agreements that allowed people to coordinate.

And coordination is everything—economically, politically, socially, even emotionally.

The problem isn't misinformation. It's infinite plausible information.

We talk a lot about misinformation as if it's the big enemy.

Misinformation is not new.

Propaganda isn't new. Scams aren't new. Manipulation isn't new.

What's new is scale.

AI doesn't just enable lies.

It enables *infinite plausible versions of reality.*

That creates a different kind of damage:

Not deception.

Confusion.

A person who is deceived can recover by discovering the truth.

A person who is confused loses the ability to *locate* truth.

Confusion is more destabilizing than deception.

Because confusion attacks the compass.

When your compass breaks, you don't just make a wrong decision.

You stop trusting yourself to make any decision at all.

That is where people become easy to manage.

Consensus weakens because shared reality weakens

Consensus requires a shared set of inputs.

Not perfect agreement—just enough overlap that people can argue within the same reality.

But in a world where:

- images can be fabricated
- voices can be cloned
- videos can be synthesized
- documents can be generated
- "experts" can be simulated
- social proof can be automated

...the shared input layer becomes unstable.

Different groups begin living inside different realities—curated by different feeds, different narratives, different AI-generated "evidence."

And because each reality is internally coherent, it feels true to the people inside it.

This is why arguments now feel less like disagreements and more like collisions between worlds.

People aren't just disagreeing about interpretation.

They're disagreeing about what happened.

The credibility crisis becomes personal

It doesn't stay out there in politics or culture.

It comes home.

Because once society stops trusting shared proxies, individuals stop trusting each other.

You see it in small interactions:

- "Did you actually say that?"
- "Is that screenshot real?"
- "Is that video edited?"
- "Did you write this message?"
- "Is this profile a person?"

Even sincerity becomes questionable.

Even apology becomes questionable.

Even warmth becomes questionable.

Not because people are colder.

Because the environment trained suspicion.

And the nervous system responds to suspicion with one of two strategies:

1. **Shutdown:** numbness, apathy, disengagement.
2. **Hypervigilance:** anxiety, scanning, compulsive verification.

Neither is freedom.

I LIVED THE HUMAN VERSION OF THIS LONG BEFORE THE TECH VERSION

This is the part where the book stays human.

Because I didn't need deepfakes to understand what it feels like when trust-by-default collapses.

I felt it the first time I realized someone could speak in a caring tone while managing me.

I felt it the first time I realized a "responsible" ritual could be a control structure.

I felt it the first time I realized an interruption wasn't random—it was timed.

I felt it the first time I noticed that I could be in a conversation for thirty minutes and then, the moment I became *me*, the room would subtly leave.

Those patterns did something to my trust.

Not my trust in "people."

My trust in my own perception.

Because if you live inside a dynamic that requires you to constantly reinterpret reality, you start to doubt the part of you that knows.

That is exactly what mass synthetic content does at scale.

It forces constant reinterpretation.

It makes you question your own eyes, your own hearing, your own memory.

It destabilizes witness.

So when I say trust-by-default ends, I'm not just talking about technology.

I'm talking about what happens to the human psyche when the environment becomes editable.

TRUST SHIFTS FROM "BELIEF" TO "VERIFICATION"

In the old world, trust was mostly belief plus history.

In the new world, trust becomes:

- verification plus pattern recognition.

This will be a hard transition because we were raised on belief.

We were taught that good people trust.

That doubting is cynical.

That skepticism is rude.

That wanting receipts is "too much."

But in a world of synthetic signal, wanting receipts is not rude.

It's sanity.

This doesn't mean you treat everyone like a criminal.

It means you stop mistaking surface credibility for inner truth.

You stop letting polished language become evidence.

You stop letting consensus become proof.

Because consensus can be engineered.

THE NEW SOCIAL TENSION: CYNICISM VS GULLIBILITY

When trust-by-default collapses, society often splits into two camps:

- the people who believe everything
- the people who believe nothing

The first group becomes easy to manipulate.
The second group becomes impossible to reach.

Both groups lose something essential:

the ability to hold reality with nuance.

Clarity is the third path.

Clarity is the ability to say:

I won't accept things just because they're presented beautifully.
And I won't reject everything just because the world is messy.

I will verify what matters.
I will watch patterns over time.
I will anchor to what is costly and consistent.

That is how you remain human in the new era.

WHY THIS IS NOT A TRAGEDY—IF YOU RESPOND CORRECTLY

The end of trust-by-default sounds bleak.

But it also creates a purification.

When cheap signal floods the world, people begin to crave costly signal again.

They begin to crave:

- honesty that isn't performative
- relationships that don't manage
- institutions that are accountable
- money that can't be edited
- leadership that is coherent under pressure

And that is why clarity becomes an asset.

Not because the world becomes worse.

Because the world becomes more persuasive.

And persuasion is not truth.

THE NEW CONTRACT

The old contract was:

If I do the right things, I'll be safe.

The new contract is:

If I keep my witness, I'll be coherent.

Safety will never be guaranteed.
Approval will never be stable.
Consensus will never be immune to manipulation.

But coherence can be cultivated.

And coherence is what lets you move through an uncertain world without becoming pliable, cynical, or lost.

When trust-by-default ends, the people who can still live are the people who can still *verify*.

Not just technically.

Spiritually.

They can say:

This is what I know.
This is what I don't know.
This is what I've seen repeat.
This is what costs something.
This is what I will no longer explain away.

That becomes the new form of trust.

Not blind trust.

Earned trust.

A ledger-based trust.

A trust that survives an editable world.

9.

The New Scarcity: Coherence

When performance becomes infinite, the world doesn't become clearer.

It becomes louder.

Not louder in volume—louder in options.

Infinite messages.
Infinite arguments.
Infinite identities.
Infinite explanations.
Infinite "clarity," delivered in perfect tone.

And when everything is available, something strange happens:

The mind stops believing in anything.

Or it starts believing in whatever feels best in the moment.

That's the fork we're approaching as technology accelerates: numbness or hypnosis.

Both are forms of captivity.

So the question becomes urgent and strangely simple:

What remains real when performance becomes infinite?

The answer is not "truth" as an abstract concept.

The answer is coherence.

COHERENCE IS NOT CHARISMA

We have confused coherence with charisma for a long time.

A charismatic person can sound coherent without being coherent.

They can speak fluently, confidently, beautifully, persuasively.

They can make the room feel organized around them.

But coherence is not the ability to organize a room.

Coherence is the ability to organize a life.

It is internal alignment:

- between what you say and what you do
- between your values and your choices
- between your feelings and your boundaries
- between your story and your ledger

Coherence is not a performance.

It is a structure.

And structures can be tested.

The old proxies are failing

In the past, we used proxies for coherence:

- credentials
- reputation
- institutional affiliation
- "expert" status
- polished speech
- the appearance of stability

These proxies were never perfect, but they often correlated with something real because it was difficult to maintain a coherent image over time without some internal stability.

Now the correlation breaks.

Because anyone can rent the image.

AI can generate the language.
Teams can generate the brand.
Platforms can generate the attention.
Networks can generate the social proof.

And once the proxies can be manufactured, we have to return to the only thing that can't be faked at scale:

Costly consistency over time.

Which is another way of saying coherence.

Coherence is costly

This is why coherence becomes scarce.

Coherence costs something.

It costs:

- saying no
- tolerating disappointment
- losing approval
- leaving a room
- being misunderstood
- grieving a fantasy
- holding a boundary even when it makes you look "difficult"
- admitting what you know without dressing it up

Performance is cheaper.

Performance can adapt to the room.

Performance can shift with the audience.

Performance can use the right words while changing nothing underneath.

Coherence cannot do that.

Coherence requires a stable internal ledger.

And a stable internal ledger requires you to stop rewriting your own experience.

THE STRONGEST SIGNAL IN THE FUTURE WILL BE A PERSON WHO CANNOT BE EDITED

This is one of the strangest things I now believe:

In the coming era, the most disruptive thing you can be is a coherent human being.

Not loud.

Not viral.

Not perfect.

Coherent.

A person whose story doesn't change based on who is listening.
A person whose boundaries don't evaporate under pressure.
A person whose values aren't a costume.
A person whose "no" is calm and final.
A person who doesn't barter truth for belonging.

That kind of person becomes rare.

Because most people will be trained by the environment to become editable:

Editable by incentives.
Editable by fear.
Editable by social pressure.
Editable by attention systems.
Editable by money.

When the world becomes editable, the coherent person becomes an anchor.

And anchors attract.

Not because people want to be controlled.

Because people are tired of drifting.

I LEARNED COHERENCE THROUGH PAIN, NOT PHILOSOPHY

I wish I could say I developed coherence through wisdom.

I didn't.

I developed it through repeated moments where my nervous system would not let me lie anymore.

Moments where I could feel the difference between being met and being managed.

Moments where I could feel the cost of performing gratitude.

Moments where I could feel myself shrinking, sentence by sentence, to protect harmony.

Moments where I could feel the quiet humiliation of being monitored under the language of responsibility.

Each time I ignored those moments, I paid.

Not in public.

Internally.

I paid with confusion.

With indecision.

With a kind of low-grade sadness that doesn't have a name because it isn't dramatic enough to be called trauma—but it is still a wound.

The wound of living out of alignment with what you know.

Coherence begins when you refuse that wound.

COHERENCE IS THE OPPOSITE OF THE "REASONABLE" SELF

My old self was a master of being reasonable.

Reasonable is adaptable.

Reasonable is socially skilled.

Reasonable is smooth.

Coherence is different.

Coherence is not primarily social.

It is internal.

Coherence is the willingness to become inconvenient to the room in order to remain true to your own witness.

This is why coherence can feel like risk at first.

Because the room often rewards reasonableness.

And the room often punishes clarity.

Especially when clarity exposes a dynamic the room depends on.

But coherence is the only thing that scales into the future without destroying you.

THE FUTURE WILL RUN ON INFINITE PERFORMANCE AND SCARCE INTEGRITY

As AI expands, more and more of life will be filled with synthetic surfaces:

- curated identities
- manufactured sincerity
- "expert" knowledge without embodied practice
- leadership theater
- empathy theater
- certainty theater
- moral theater

This isn't a moral panic.

It's a predictable outcome of abundant language.

When the supply of persuasive words goes up, people will use persuasive words.

They will use them to win.

To sell.

To manage.

To soothe.

To control.

What will be rare is the person who doesn't need to perform.

The person who can be plain.

The person who can say:

I don't know.
I was wrong.
This is what happened.
This is what it cost.
This is what I will not do again.

That person will feel almost sacred in the new environment.

Because sacredness is not mysticism.

Sacredness is rarity plus truth.

COHERENCE IS THE NEW WEALTH

This is where the idea of "clarity as the best asset" becomes more than a metaphor.

In the old world, people tried to acquire wealth to feel safe.

In the new world, wealth without coherence will not protect you.

Because without coherence, you can be persuaded into bad decisions.

You can be manipulated by narratives.

You can be sold counterfeit certainty.

You can be made pliable.

So coherence becomes a form of wealth:

- it protects your attention

- it protects your boundaries
- it protects your money decisions
- it protects your relationships
- it protects your identity
- it protects your peace

And like wealth, coherence compounds.

The more coherent you become, the less you leak.

The less you leak, the more power you have.

Not power over others.

Power over your own life.

THE TELL: COHERENCE SHOWS UP UNDER PRESSURE

You can fake almost anything in a calm environment.

You can fake kindness.
You can fake principles.
You can fake clarity.

But you cannot fake coherence under pressure.

Pressure reveals whether the ledger matches the language.

When a person is stressed, do they become honest or do they become manipulative?
When they are challenged, do they become accountable or do they rewrite the past?
When they lose control, do they become truthful or do they become strategic?

Thc samc is true for institutions.

The same is true for systems.

Pressure is the audit.

In an age of infinite performance, pressure is where reality returns.

The simplest definition

If I had to define coherence in one line, it would be this:

Coherence is living in a way that your future self will not have to explain.

No elaborate justifications.
No revisionist memory.
No story that changes to protect your pride.

Just clean truth.

That's the new scarcity.

Not content.

Not information.

Not intelligence.

Coherence.

And the person who builds it now—quietly, deliberately—will be difficult to manage later.

Because they will have something that the future cannot generate:

A life that matches its words.

A ledger that holds.

A self that stays.

Tech Interlude C

Proof of Work vs Proof of Words

The world is about to be flooded with words that sound like proof.

Perfectly reasoned arguments.
Beautifully balanced perspectives.
Sincere-sounding apologies.
Deep-sounding insight.
Warm, "human" messages that feel intimate.

And the danger won't be that the words are always false.

The danger will be that the words are no longer *evidence* of anything.

Because once AI can generate language on demand, language stops functioning as a reliable signal of inner life.

That's why we need a new distinction:

Proof of words vs proof of work.

Proof of words: the easy signal

Proof of words is what most of modern culture runs on already:

- "I care about people."
- "We value transparency."
- "I'm committed to growth."
- "Safety is our top priority."
- "This is best practice."
- "Trust us."

These phrases can be true.

But they can also be manufactured.

And increasingly, they will be manufactured—at scale, with flawless tone.

AI makes proof of words abundant:

- You can generate a mission statement in seconds.
- You can generate a heartfelt message in seconds.
- You can generate a nuanced position in seconds.
- You can generate a persuasive explanation in seconds.

Words become low-cost output.

Which means words become low-cost signal.

And low-cost signal gets noisy.

Proof of Work: The Costly Signal

Proof of work is different.

Proof of work is what cannot be produced without cost.

It requires:

- time
- attention
- discipline
- sacrifice
- consistency
- follow-through
- accountability
- risk

It shows up as outcomes, not claims.

Not "I care about people," but *how you treat people when they are inconvenient.*

Not "We value transparency," but *whether your system can be audited.*

Not "I'm committed," but *what you do when no one is watching.*

Proof of work is inherently harder to fake because it leaves a trail.

It accumulates.

It creates a pattern.

And patterns are more difficult to manufacture than sentences.

Proof of Work (Conceptual)

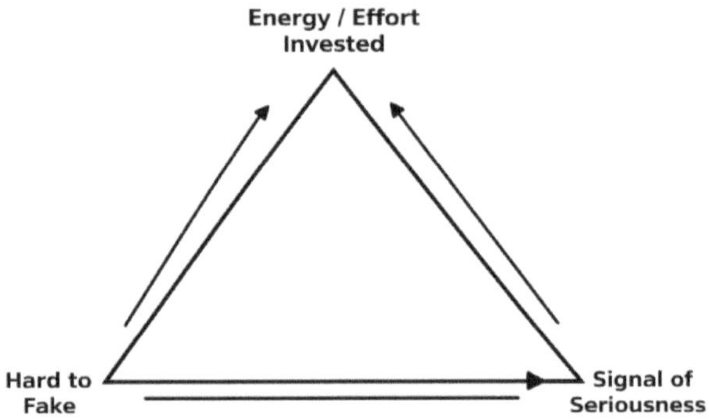

Work is credibility you can't copy/paste.

THE CORE TECHNOLOGY CONCEPT

Here is the simplest way to understand it:

AI produces **signal** without cost.

It can mimic the outward markers of intelligence, empathy, clarity, competence—because it can generate the *appearance* of those things.

But only work produces **costly signal**.

Costly signal is credible because it is expensive to maintain over time without actually having the underlying structure.

This is an old truth in nature and human life.

A peacock's feathers are costly.
A long apprenticeship is costly.
A reputation built over decades is costly.
A consistent pattern of integrity is costly.

Cost filters out most pretenders.

That's why cost matters.

And that's why the era we're entering changes the rules: it removes cost from the most persuasive form of human signal—language.

Why "clarity" becomes confusing in the AI era

Clarity has always had two versions:

1. Clarity as language — "I can explain this well."
2. Clarity as coherence — "I know what is true and I live from it."

In the AI era, version (1) becomes cheap.

Anyone can sound clear.

But version (2) becomes rare.

Because coherence still costs something.

It costs refusing to rewrite your experience.
It costs holding your boundary.
It costs leaving misaligned rooms.
It costs choosing truth over belonging.
It costs accepting short-term discomfort for long-term integrity.

AI can generate the words of clarity.

It cannot generate the cost of coherence.

That's why clarity is the first asset: it is proof of work inside the self.

Proof of work in human systems

I know this from my professional life.

In complex transitions—carve-outs, integrations, messy operational realities—language is cheap.

Everyone has a slide deck.

Everyone has a "plan."

Everyone has a narrative about what is happening.

But the truth is always in the work.

In who actually owns the problem when it's ugly.

In who follows through when the excitement is gone.

In who keeps the record clean when nobody's looking.

In who tells the truth when it costs them politically.

In those environments, you learn quickly:

You can't run a project on words.

You run it on proof.

And as the world becomes saturated with synthetic signal, that same rule will govern human life.

How You Tell the Difference in the New Era

In a persuasion-rich world, the main question shifts from:

"Does this sound right?"

to:

"What does this cost?"

Cost doesn't only mean money.

Cost means:

- What would it cost this person to maintain this story consistently?
- What would it cost this institution to be accountable?
- What would it cost this platform to tell the truth?

- What would it cost this relationship to meet me instead of manage me?

This is why clarity is not just a personal virtue.

It's an analytical tool.

Clarity is the ability to see cost, patterns, incentives—and not be hypnotized by polished speech.

Why the ledger matters here

Proof of work requires a ledger.

Not necessarily a public ledger.

A record.

A way to track what is true over time.

Because proof of words is momentary.

A single message can be perfect. A single apology can be flawless. A single post can be convincing.

But a ledger forces continuity.

A ledger reveals whether the pattern matches the performance.

That's why ledgers—personal and technological—matter in the world we're entering.

They anchor reality to record.

They make revision harder.

They make manipulation more visible.

The transition we're living through

We are moving from a world where words often correlated with reality…

to a world where words will often correlate with optimization.

Optimization for attention.
Optimization for compliance.
Optimization for persuasion.
Optimization for conversion.

So the human advantage shifts.

The advantage becomes:

- clarity that can't be prompted
- integrity that can't be generated
- proof of work that can't be simulated over time
- coherence that survives pressure

This is why I keep returning to the same claim:

Clarity is the first asset.

Because in the end, your clarity is not what you can say.

Your clarity is what you can hold.

And what you can hold—consistently, under pressure—becomes the only signal that matters when everything else becomes infinite.

ACT IV

THE RECKONING

10.

Proof of Work in Human Systems

If you want to understand what proof of work looks like in human life, don't look at speeches.

Look at transitions.

Look at the places where the glossy story ends and the actual work begins—where the power dynamics surface, where the incentives reveal themselves, where people discover what they really value by what they abandon.

In my world, that place has a name.

The messy middle.

It's the stretch of time after a deal closes and before anything truly works. The stretch where institutions are still performing confidence while the ground underneath them shifts. The stretch where people say "alignment" but mean "control," where they say "process" but mean "delay," where they say "best practice" but mean "we don't want liability."

It's the stretch where there are no applause lines.

Just work.

And I have spent years living there.

THE MYTH OF THE PLAN

Every transition begins with a plan.

A deck. A roadmap. A timeline with clean arrows and neat boxes.

People gather in meetings and talk about:

- governance
- milestones
- stakeholders
- operating models
- value creation
- integration strategy

It sounds coherent.

It often is coherent—on paper.

Then reality arrives.

Reality arrives in small ways at first:

- contracts that don't match the narrative
- vendors who don't agree with what "the plan" assumes
- data that is incomplete

- systems that don't talk to each other
- legal obligations that were minimized in the story
- people who are quietly leaving
- teams that are exhausted and already under-resourced

And this is where the real truth of the deal begins to appear:

The deal was not just a financial event.

It was a human event.

And humans cannot be moved on a slide.

THE MESSY MIDDLE IS WHERE INCENTIVES BECOME VISIBLE

In the messy middle, you learn quickly that people don't optimize for what they say they optimize for.

They optimize for what costs them personally.

A leader may say they want transparency, but if transparency threatens reputation, transparency dies.

A team may say they care about operational rigor, but if rigor slows the timeline, rigor gets cut.

A company may say "people first," but if "people first" affects EBITDA, people become a line item.

This is not cynicism.

This is literacy.

It is the ability to see incentives without needing to hate anyone for having them.

Incentives explain behavior far more reliably than values statements do.

And once you begin to see incentives clearly, you stop expecting the world to behave like a moral essay.

You start building systems that can survive the truth.

STAYING WHEN NO ONE CLAPS

There is a moment in every transition—often many moments—where the work becomes unglamorous.

It's not strategic anymore.

It's operational.

It's detail.

It's follow-up.

It's the long, quiet labor of making the story true.

That's the moment most people detach.

Not because they're lazy.

Because they're rewarded for visibility, not completion.

In corporate environments, you are often rewarded for:

- presenting the plan
- attending the meeting
- signaling alignment
- speaking the language
- being seen in the room

But you are rarely rewarded for:

- holding thc thread
- owning the ugly parts

- following the work through to done
- telling the uncomfortable truth early enough to prevent disaster
- being the person who keeps the record clean

That kind of work is invisible until it fails.

And because it's invisible, it doesn't come with applause.

So you either develop a taste for internal integrity—or you drift toward performance.

I learned to anchor to integrity because I had to.

In the messy middle, performance doesn't protect you.

It collapses.

The Ledger as Operational Truth

This is where my love of the ledger stops being metaphor.

In transitions, the ledger is what saves you.

Not always a literal ledger, but a record that can't be argued with:

- contract inventories
- obligation tracking
- issue logs
- decision records
- vendor communications
- milestones
- risk registers
- handoff documentation

In a carve-out or integration, memory is not enough.

Tone is not enough.

Intent is not enough.

Even trust is not enough.

Because teams change, leadership shifts, priorities pivot, and the story gets rewritten in real time.

So you need something stable.

You need a record that can hold reality when people can't.

And once you've built that kind of record professionally, it becomes impossible not to see where your personal life lacks it.

If you are living inside a relationship or an institution that depends on your forgetting—your forgetting what happened, what was said, what it cost—you are in an editable system.

And editable systems always favor the person with more power.

So I began to treat my personal life like I treated complex projects:

Not as a performance.

As a set of truths that needed to be recorded.

Not because I was keeping score.

Because I was keeping sanity.

THE DIFFERENCE BETWEEN TALKERS AND BUILDERS

Over time, you learn to recognize two types of people.

Talkers.

Builders.

Talkers are often talented. They have language. They have charm. They can create momentum in a room. They can make the plan sound inevitable.

They can make you feel like progress is happening even when nothing changes.

Builders are different.

Builders are often quieter.

They may not dominate meetings. They may not be praised. They may not even be liked.

But they do something rare:

They take responsibility for what is true.

They don't disappear when things get hard.

They don't outsource the messy parts.

They stay.

And staying is a form of proof.

Staying doesn't mean staying in bad situations, staying in abusive dynamics, staying in misalignment.

Staying means staying with the work when it loses glamour.

Staying means not abandoning reality when reality becomes inconvenient.

That is proof of work.

PROOF OF WORK IS THE ONLY CURRENCY THAT HOLDS THROUGH CHAOS

In a transition, chaos is not an exception. It's the environment.

So you discover quickly what holds:

- a clear record
- a coherent operating model
- people who tell the truth early

- follow-through
- ownership
- accountability
- the willingness to do what can't be spun

Everything else—tone, brand, presentation—can help temporarily.

But when pressure comes, performance collapses.

What remains is what was built.

This is why I began to think about proof of work not as a tech phrase, but as a human necessity.

If you can't distinguish proof of words from proof of work, you will be managed by speech.

In business.

In institutions.

In relationships.

And in the era we're moving into—where words can be generated endlessly—you will be at the mercy of whoever can produce the most convincing narrative.

So proof of work becomes your anchor.

THE RECKONING: CHOOSING WHAT I ANCHOR TO

Act IV is where you stop hoping the world will become simpler.

You accept that complexity is the truth.

And then you choose what you will anchor to inside that complexity.

For me, the anchor became obvious:

- I anchor to record over narrative.

- I anchor to pattern over promises.
- I anchor to cost over tone.
- I anchor to follow-through over performance.

That anchoring didn't make me harder.

It made me clearer.

It made me less susceptible to being hypnotized by competent speech.

It made me less willing to participate in systems that depended on my forgetfulness.

Because once you've lived through enough messy middles, you know the most important truth of all:

What matters is not what people say in the beginning.

What matters is what they do when the work is ugly and no one is clapping.

That is where character is revealed.

That is where institutions show their real incentives.

That is where your own clarity is tested.

And that is where the ledger stops being a concept and becomes a practice:

You choose what is true.

You record it.

You build from it.

You stop living in editable reality.

You stop being managed by words.

You become a builder of something harder—something that can survive pressure.

A life anchored to proof.

11.

Money as a Ledger

For most of my adult life, I thought money was the practical part of life.

The necessary part.
The boring part.
The part you handle so you can get back to what's real—love, work, family, meaning.

But over time I realized money isn't just practical.

Money is philosophical.

Money is relational.

Money is a system of truth—because money is a ledger.

And once you understand money as a ledger, you start noticing how often the ledger can be edited.

Not only by governments.

By institutions.
By policies.
By gatekeepers.
By "risk departments."
By social dynamics inside relationships.

That's when money stops being boring.

That's when money becomes a clarity problem.

Permission money: the ledger you don't control

Most of the money we use is permission-based.

That doesn't mean it's evil.

It means it functions through gates.

A bank account is permission-based.
A credit line is permission-based.
A payment platform is permission-based.
A card transaction is permission-based.

You are allowed to transact because you remain inside the system's rules, and the system remains willing to serve you.

There's nothing inherently wrong with this—until you realize what permission does to the psyche.

Permission money subtly trains you to live as a guest in your own life.

It trains you to behave.

To remain legible.
To remain acceptable.
To remain low-risk.
To remain compliant.

And in exchange, you get access.

Access to your own funds.
Access to payments.
Access to the ability to move value through the world.

In good times, permission feels invisible.

In stressful times, permission reveals itself.

A hold.
A delay.
A "review."
A frozen transfer.
A closed account.
A request for documentation that arrives like a threat.

And the terrifying part is not the friction.

The terrifying part is the silence.

Because permission money doesn't have to explain itself to you in a way you can audit.

It just has to protect the institution.

I LEARNED PERMISSION INSIDE A MARRIAGE BEFORE I LEARNED IT IN FINANCE

This is where the personal and the systemic converge.

In a relationship where money is used as management, you learn permission before you ever learn macroeconomics.

You learn what it feels like to justify.
To account.
To prove you're responsible.
To be evaluated.

Weekly budget meetings that look "responsible" can become a ritual of hierarchy.

Not always intentionally. But structurally.

And even when the facts don't support the posture—even when you earn more, even when you're steady—the ritual creates a subtle message:

Someone is monitoring. Someone is authorizing. Someone is in charge.

That dynamic teaches you a lesson that goes deeper than dollars:

Money isn't only a tool.

Money is a lever.

Money can be used to shape your behavior.

So when I later began thinking about money systems, I wasn't starting from theory. I was starting from lived sensation:

The feeling of being permitted.

The feeling of needing approval to access what should already be yours.

VERIFIABLE MONEY: THE LEDGER YOU CAN AUDIT

Then I encountered something that landed differently—not as hype, but as relief.

A ledger that could be verified.

A system where the rules were explicit and the record was public.

A form of money that didn't rely on an institution's willingness to serve you.

This is where "hard truth" enters the picture.

Hard truth isn't harshness.

Hard truth is simply reality that cannot be rewritten because someone is uncomfortable.

Hard truth is reality that holds even when the story changes.

Hard truth is what remains when performance collapses.

And that's what a verifiable ledger represents to me: a kind of truth that is structural, not emotional.

In a world where so much is editable—where words can be generated, narratives can be managed, access can be filtered—something that can be verified feels like a lifeline.

Not because it makes life perfect.

Because it makes reality stable enough to build on.

WHY I MOVED TOWARD "HARDNESS"

It's easy for people to misunderstand why someone would move toward hard money.

They think it's greed.

Or politics.

Or fear.

For me, it was clarity.

It was the recognition that softness has a cost.

Soft money can be expanded.
Soft rules can be changed.
Soft access can be revoked.
Soft narratives can be rewritten.

Softness can feel compassionate, but it can also be a tool for control.

Because if the ledger can be edited, the person who can edit it has power over everyone who depends on it.

So "hardness" isn't about becoming cold.

It's about removing discretionary power from places where discretion can be abused.

Hard truth reduces the need to trust.

And in the age we're entering—where trust-by-default collapses—that matters.

Permission vs Verification: A Human Distinction

The most important distinction isn't technical.

It's human.

Permission money asks:

Will you behave? Will you comply? Will you remain acceptable?

Verifiable money asks:

Can you prove it? Can you verify it? Can you audit it?

The first is relational—dependent on gatekeepers.
The second is structural—dependent on rules and record.

This is why the ledger concept keeps returning. It shows up everywhere:

In relationships: do you need permission to be heard?
In institutions: do you need permission to be recognized?
In money: do you need permission to transact?
In the future: do you need permission to remain legible?

When you live long enough inside permission, verification feels like oxygen.

The Future Will Blur Money and Identity

This is the part that makes clarity urgent.

Money systems are increasingly tied to identity systems.

Not only in obvious ways like KYC, but in subtle ways:

risk scoring, fraud models, "trust" metrics, behavioral monitoring.

Whether you call it surveillance or compliance depends on your politics.

But the lived experience is similar:

Access becomes conditional.

And conditions are rarely neutral.

They are shaped by institutional incentives: to reduce risk, reduce liability, reduce exposure.

This means the future will likely offer money that is increasingly "smart" at the edges—more controllable, more programmable, more filterable.

Even if the stated intent is safety, the human effect can be constriction:

- delays
- holds
- restrictions
- approvals
- denials
- "We can't explain this decision"

So the question becomes:

What kind of money do you want to hold when the rails become more conditional?

WHY HOLDING A HARD TRUTH CHANGES YOUR PSYCHOLOGY

This is something people don't talk about enough: money changes your nervous system.

Not because of greed.

Because money is time.

Because money is options.

Because money is the ability to leave.

When you hold money that can be frozen, your nervous system knows it—even if you don't think about it daily.

When you hold money that is harder to seize, harder to rewrite, harder to debase, your nervous system feels something else:

Stability.

Not guaranteed safety—nothing guarantees safety.

But stability enough to make clean decisions.

And clean decisions are the whole theme of this book.

Clarity is the asset because clarity is what keeps you from being managed.

Hard truth is attractive because it reduces management.

It reduces the places where you have to beg, plead, explain, perform.

It reduces the number of rooms where your reality can be overwritten.

My reason was not ideology. It was alignment.

I didn't move toward verifiable money because I wanted to be radical.

I moved toward it because I was tired of living in systems where:

- the story changes
- the rules shift
- the ledger is edited
- the gatekeeper is invisible
- and the burden of proof is always on you

I had spent years building ledgers professionally so that transitions could survive reality.

At some point, I realized I wanted the same thing personally:

A ledger that holds.

A form of truth that doesn't depend on anyone's mood, or narrative, or permission.

The real point

This isn't a claim that verifiable money solves human suffering.

It doesn't.

Human suffering finds new shapes in every era.

The point is simpler and more humane:

When the world becomes more editable, you will crave something that isn't.

And money is one of the most fundamental places to choose what is editable and what is not.

Because you can't build a coherent life on a ledger that is constantly being rewritten by forces you can't see.

So I began to anchor to harder truth—not out of fear.

Out of respect for reality.

Out of the same instinct that made me stop explaining.

Out of the same instinct that made my internal ledger appear.

If clarity is the first asset, then money is one of the first places you either protect it or lose it.

Permission money trains you to be manageable.

Verifiable money trains you to be accountable—to reality, not to gatekeepers.

And in the end, what I wanted was not control over others.

It was something much quieter:

To live in a world where my own life could not be rewritten without my consent.

To hold a ledger that matched the truth I was finally willing to live.

The Ledger Principle

Without verification. trust becornes theater.

12.

Institutions as Machines

It took me a long time to stop taking institutions personally.

When an institution fails you, the first instinct is to interpret it as judgment:

I did something wrong.
I'm not enough.
I didn't follow the process.
I should have been clearer.
I should have known.

That instinct is deeply human. We want the world to be relational because we are relational. We want the system to behave like a person—fair, responsive, accountable, capable of listening.

But institutions are not people.

They are machines.

And once you see them as machines, the pain doesn't vanish—but the confusion does.

Because you stop asking a machine to love you.

INSTITUTIONS ARE BUILT TO SCALE, NOT TO WITNESS

A human can witness you.

A human can hold nuance, remember context, track emotional truth, and respond with care.

An institution cannot do that at scale.

To scale, it must reduce complexity into categories.

It must standardize.

It must decide what counts and what doesn't.

So it builds interfaces—forms, portals, policies, procedures—that translate living reality into legible input.

That translation is inherently violent in a small, quiet way.

It cuts off nuance.

It trims the edges.

It compresses a life into something the machine can process.

And once your life becomes processable, it becomes governable.

This is what people mean—whether they know it or not—when they talk about bureaucracy.

Bureaucracy is not just paperwork.

It's reality compression.

WHAT INSTITUTIONS OPTIMIZE FOR

Institutions tell you what they value.

They have posters and mission statements and leadership speeches.

But institutions don't run on posters.

They run on incentives.

And incentives reveal optimization.

In my work, I have watched enough complex organizations under pressure to name the most common optimizations clearly:

1) Liability reduction
Institutions optimize to avoid blame.

If a choice increases the chance of being sued, criticized, audited, or punished publicly, it will be resisted—even if it's humane.

2) Risk containment
Institutions optimize to prevent outliers.

Outliers create unpredictability. Unpredictability creates exposure. Exposure creates cost.

So institutions create policies that treat outliers as threats.

3) Reputation management
Institutions optimize to protect the narrative.

Not necessarily to lie—sometimes simply to control the framing.

Acknowledge only what is necessary. Admit only what is survivable.

4) Budget and resource efficiency
Institutions optimize for cost control.

Even when the institution says it values people, it will still choose efficiency over care when under strain.

5) Continuity of power
Institutions optimize to preserve hierarchy.

Decisions that concentrate authority tend to persist. Decisions that decentralize authority tend to meet resistance.

6) Compliance with the institution's own rules
Institutions optimize for procedural correctness.

Not moral correctness—procedural correctness.

This is why a human can be obviously harmed while the institution says, "We followed policy."

To a machine, following policy *is* correctness.

This list isn't cynical. It's just accurate.

And accuracy is the beginning of clarity.

INSTITUTIONS CAN BE STAFFED BY GOOD PEOPLE AND STILL BEHAVE INHUMANELY

This is one of the hardest truths to accept.

You can have kind people inside a machine—and the machine can still produce cold outcomes.

Because individual virtue does not override structural incentives.

A good person can fight the machine, but they often do so at personal cost.

They may be punished for it.
They may be exhausted by it.
They may eventually leave.

So over time, the machine selects for people who can tolerate its logic.

Not always the worst people.

Often the most adaptive.

The ones who can translate inhuman outcomes into reasonable language.

The ones who can say, with calm sincerity:

This is for safety.
This is policy.
This is best practice.
This is how the system works.

Which brings us back to the era we're entering: proof of words becomes abundant, and institutions get better at sounding humane without changing their incentives.

What institutions can never provide

Here is the line I had to cross internally:

There are certain things an institution cannot provide, no matter how well designed it is.

It cannot provide:

1) Love
It can provide benefits. Services. Programs. But it cannot provide love.

Love requires a face. Love requires presence. Love requires witnessing.

2) Meaning
Institutions can provide frameworks, but meaning is an internal construction.

If you ask an institution to give you meaning, you will become manageable.

3) Identity
Institutions can issue credentials and IDs, but they cannot tell you who you are.

If you let them, you will live in permission.

4) Dignity
Institutions can treat you politely, but dignity is not something a machine can grant.

Dignity is something you protect by refusing to cooperate with your own diminishment.

5) True safety

Institutions can reduce certain risks, but they cannot guarantee safety.

If you believe they can, you will trade your autonomy for the illusion of protection.

This is the old contract again:

If I do the right things, I'll be safe.

Institutions sell that contract because it keeps people compliant.

But it isn't true.

THE HIDDEN TRADE: SECURITY FOR COHERENCE

When people are afraid, they ask institutions to manage uncertainty.

That's understandable.

But the trade is often deeper than people realize:

You trade your ability to decide for yourself in exchange for a system that claims it will protect you.

Sometimes that trade is worth it.

But when it becomes unconscious—when you stop noticing how much you are surrendering—you become a person shaped by permission.

That is the spiritual danger of institutional dependence.

Not that institutions exist.

That you start using them as your internal compass.

I LEARNED THIS IN THE MESSY MIDDLE

In transitions, you see the machine clearly because everything is stressed.

Stress reveals optimization.

When a deal is under pressure and timelines are tight and leaders are exposed, the institution's real values show up quickly.

Not in speeches.

In actions.

You see what gets prioritized, what gets delayed, what gets denied, what gets quietly buried.

You watch how fast "people first" becomes "risk first."

You watch how often "transparency" becomes "controlled messaging."

You watch how quickly the system rewards the person who manages perception more than the person who fixes reality.

And you learn: the machine does what it is designed to do.

That's not personal.

It's architecture.

PAST VS FUTURE: INSTITUTIONS BECOME MORE AUTOMATED, LESS RELATIONAL

In the past, an institution still had more human seams.

You could find a person. You could call someone. You could appeal.

In the future, the machine becomes cleaner and more automated.

More decisions get routed through:

- scoring models
- policy engines

- risk algorithms
- automated holds
- template communications
- invisible moderation
- system-driven eligibility

This increases efficiency.

It also increases opacity.

And the more opaque the machine becomes, the more people feel:

I'm being acted on by something I cannot speak to.

That feeling is what collapses trust.

It's what makes people either cynical or pliable.

So this chapter matters because it's not only analysis—it's a preparation.

If you understand institutions as machines, you stop expecting them to meet you.

You stop begging the machine to be human.

And then you can do something freeing:

You can choose your anchors.

THE RECKONING: CHOOSING WHAT I ANCHOR TO

Act IV is where you choose what you will build your life on.

Not ideals.

Anchors.

For me, the anchors became clear:

- I anchor to my internal ledger, not institutional recognition.

- I anchor to verifiable truth where possible, not narrative assurance.
- I anchor to relationships that meet me, not systems that manage me.
- I anchor to coherence over compliance.

This doesn't mean rejecting institutions.

It means placing them in their proper role.

Institutions can support life.

They cannot be life.

They can be tools.

They cannot be your witness.

They can issue forms.

They cannot tell you what is true.

And once you stop asking the machine to provide what it cannot provide, something quiet happens:

You become harder to manipulate.

Not because you become distrustful.

Because you become clear.

You stop seeking permission for your own reality.

You stop waiting for the system to validate what you already know.

You stop confusing recognition with truth.

That is the reckoning.

And it is also the beginning of sovereignty.

Tech Interlude D

The Ledger War

There's a quiet tug-of-war happening underneath modern life.

Most people feel it as mood—unease, fatigue, suspicion, numbness—without knowing what they're sensing.

But if you zoom out, it's structural:

What kind of ledger will the world run on?
And who gets to edit it?

I'm calling this a "ledger war," not because it's dramatic, but because it is a real tension—an ongoing negotiation between two directions the future could take, often at the same time.

And the most important thing to say up front is this:

This is not a conspiracy story.
It's not a panic story.
It's not even a politics story.

It's an incentives story.

TWO MOVEMENTS CAN BE TRUE AT ONCE

We keep wanting a single narrative:

Either the world is decentralizing…
or the world is centralizing.

But reality is more complex.

Two movements are happening simultaneously:

1. **Decentralizing base layers** — systems designed to be open, auditable, hard to edit.
2. **Increasingly controlled rails** — systems designed to manage access, manage risk, manage compliance, manage outcomes.

Both are expanding.

And they expand for understandable reasons.

Decentralizing base layers expand because trust is weakening and people want verifiable truth.

Controlled rails expand because institutions have obligations—fraud prevention, consumer protection, sanctions, AML compliance, systemic stability, liability reduction.

Neither movement is pure good or pure evil.

They simply optimize for different things.

And optimization always creates tradeoffs.

BASE LAYERS VS RAILS: A CLEAN DISTINCTION

Here's the simplest way to hold it:

Base layer = the ground truth record.
Rails = the pathways that let most people interact with that record.

Think of the internet.

The base protocol might be open, but most people experience the internet through platforms—controlled layers that shape what you see and what you can do.

Money and identity are moving the same way.

A decentralized ledger can exist as a base layer—auditable, verifiable, difficult to edit.

But most people won't interact with it directly.

They'll interact through rails:

- exchanges
- banks
- payment apps
- custody providers
- compliance gateways
- regulated intermediaries
- user-friendly platforms

Rails add usability.

Rails also add control.

Because rails have gatekeepers.

Gatekeepers don't always act maliciously.

They act according to incentives and obligations.

And the incentives of rails tend to favor manageability.

WHY "CONTROLLED RAILS" KEEP GROWING (EVEN WITH DECENTRALIZATION)

This part is not scary—it's predictable.

As systems scale, institutions:

- must reduce fraud
- must comply with law
- must manage liability
- must standardize experiences
- must prevent outliers from breaking the system
- must protect customers (and themselves)

So controlled rails proliferate because:

the world is demanding safety and convenience at scale.

And safety + convenience usually require:

- monitoring
- scoring
- identity binding
- permissioning
- reversibility
- policy enforcement

Those are not necessarily "evil." They are simply the tools of institutional risk management.

But those tools create a lived experience:

Access becomes conditional.
Transactions become filterable.
Accounts can be paused, reviewed, denied.
Explanations become templated or opaque.

This is the new "form." This is the new gate.

Why decentralizing base layers keep growing

At the same time, people are craving something stable beneath the noise.

A record that can't be rewritten by a committee or updated away by a policy change.

A ledger that can be audited.

A ground truth that doesn't depend on mood, narrative, or permission.

This grows because:

trust-by-default is ending (as we discussed) and because the environment is becoming synthetic.

When words become cheap and images become questionable, the desire for verifiable anchors intensifies.

So the base layer movement isn't mainly ideological.

It's psychological.

It's the human need for reality to hold.

The actual "war" is a negotiation over editability

If you remove the drama, the tension comes down to one question:

Where will editability be allowed?

- Will the ledger itself be editable?
- Or will the ledger be hard, while rails are editable?

- Will rails be lightly permissioned, or heavily permissioned?
- Will identity be separable from transactions, or fused to them?
- Will verification be user-controlled, or institution-controlled?

Different societies, industries, and companies will answer differently.

There will be hybrid systems.

There will be cycles.

There will be compromises.

That's not fear. That's how systems evolve.

WHERE PEOPLE GET LOST: TURNING IT INTO A MORAL IDENTITY

This is where I want to be careful.

Because it's easy to turn the ledger war into a personality type:

- "decentralized people" vs "centralized people"
- "freedom people" vs "control people"
- "Bitcoin people" vs "institution people"

That framing is emotionally satisfying.

It's also too simple.

The truth is: most humans want both.

They want autonomy *and* safety.
They want privacy *and* protection.
They want freedom *and* stability.
They want verification *and* convenience.

So the question isn't: which side are you on?

The question is: *what are you anchoring to when the tradeoffs appear?*

My point: clarity isn't a political stance—it's orientation

Clarity is not a flag you wave.

Clarity is an internal posture:

- I can see incentives.
- I can see tradeoffs.
- I can see what is being optimized for.
- I can see where editability lives in the stack.
- I can choose consciously rather than drift.

This is the real "upgrade" for the coming era.

Because the danger isn't that controlled rails exist.

The danger is living unconsciously inside controlled rails while telling yourself you are free.

The danger is letting convenience become your only compass.

Clarity is what prevents that.

Clarity lets you say:

- *This rail is convenient, and it comes with conditions.*
- *This base layer is verifiable, and it comes with responsibility.*
- *This institution reduces certain risks, and introduces others.*
- *This technology increases freedom in one layer and reduces it in another.*

Clarity is how you stay sane in a world of mixed truths.

The personal echo: I've lived "rails" in relationships

This is why I can't separate the technology story from my human story.

Because I've lived this tension at the emotional level:

Being "met" is like open protocol—presence, witness, truth.
Being "managed" is like controlled rails—polite, efficient, conditional.

When I finally saw the difference, I stopped moralizing it.

I stopped expecting managed systems to behave like human witness.

I started choosing my anchors.

That's what this interlude is really saying:

Don't panic. Don't romanticize. Don't outsource your mind.

Just see clearly.

The Practical Takeaway

If the world is moving toward a hybrid stack—harder base layers + more managed rails—then your job is not to pick a team.

Your job is to become literate.

To know where you're standing.

To know what layer you're using.

To know what you're trading.

And to keep an internal ledger that protects your coherence when external ledgers become contested.

Because the ledger war is not only out there.

It's inside every decision you make about:

- where you store value
- how you identify yourself
- what conveniences you accept
- what permissions you trade for speed

- what truths you anchor to when narratives shift

Clarity isn't a stance.

It's orientation.

And orientation is what keeps you from being moved—by fear, by hype, by rhetoric, by perfect words—when the stack gets more complex.

That is the whole point of this book.

ACT V

THE ERA OF SOVEREIGNTY

13.

Clarity in Love

Love is where clarity becomes unbearable.

Because you can be unclear in many areas of life and still function.

You can be unclear in career for a while and still pay your bills.
You can be unclear in identity for a season and still move forward.
You can be unclear about money and still get by.

But love is different.

Love exposes every place you are negotiating with yourself.

Love is where you discover what you actually believe you deserve—not in theory, but in behavior.

And for me, the greatest clarity in love came down to a simple distinction:

Met vs managed.

Met

To be met is to feel the other person's presence arrive.

Not their opinion.

Not their advice.

Not their performance.

Their presence.

Being met feels like:

- they stay with your sentence
- they hold the thread
- they don't compete with your joy
- they don't redirect when you become alive
- they can witness you without making it about them
- they are curious, not threatened
- they respond to your interior world as if it matters

Being met doesn't mean being agreed with.

It means being received.

It's the feeling of: *I exist here. I am safe to be real.*

Managed

To be managed is to feel your reality being regulated.

Not always harshly.

Often politely.

Often invisibly.

Being managed looks like:

- interruption right when you become bright
- subject change that lands like a quiet erasure
- "helpful" advice that bypasses your truth
- jokes that soften your confidence
- budget meetings that feel like audits
- gifts that carry a subtle hierarchy
- affection that arrives when you are easy and recedes when you are fully you

Management doesn't always look like control.

Sometimes it looks like being tolerated.

Sometimes it looks like being "handled" with care.

But the internal effect is consistent:

You begin to edit yourself.

And the moment you start editing yourself, love becomes conditional.

Not necessarily conditional in their words.

Conditional in your nervous system.

Because your body learns:

My aliveness has a cost.

JOY AS THE TEST

This is the part that changed my life.

I used to think love was tested by hardship.

By who shows up when things are bad.

That matters.

But there's another test that is even more revealing:

Who can be with your joy?

Not your crisis.

Your joy.

Your excitement.
Your pride.
Your creative energy.
Your momentum.
Your becoming.

Joy is strangely confronting to many people.

Because your joy contains information:

- that you are growing
- that you don't need them to be okay
- that you have an inner world that is self-sustaining
- that you are not manageable through scarcity
- that you are becoming more yourself

Joy is not needy.

Joy is sovereign.

And sovereign energy can scare people who unconsciously rely on you being smaller.

So the joy test is brutal in its honesty:

When I light up, do you lean in or do you leave?

Do you become curious or do you interrupt?

Do you celebrate or do you regulate?

Do you meet me or manage me?

This is why I started noticing interruptions differently—not as small social mishaps, but as data points in the love ledger.

Because if your joy consistently triggers withdrawal, you are not being loved in a way that can hold your full self.

You are being loved in a way that can hold your edited self.

THE REAL GRIEF: LEAVING THE FANTASY OF BEING MET WHERE YOU ARE NOT

The hardest part of clarity in love is not identifying what is happening.

The hardest part is grieving what will not happen.

Grieving the fantasy that if you become more reasonable, more polished, more patient, more strategic, you will finally be received.

Grieving the fantasy that the next conversation will be different.

Grieving the fantasy that the person will suddenly be able to witness you.

This grief is clean.

It's not dramatic.

It is simply the ending of a hope that kept you in negotiation with yourself.

And when that hope ends, something else begins:

You stop offering your deepest joy to rooms that cannot hold it.

You stop performing your smallness as a peacekeeping gesture.

You stop paying for belonging with self-erasure.

That is sovereignty in love.

PAST VS FUTURE: "MET" BECOMES A RARE LUXURY

In the past, being met was easier because life was slower.

There were more human seams.

Fewer distractions.

Less synthetic signal.

A person's attention still had weight.

Now attention is being harvested.

Competed for.

Fragmented.

Optimized.

Which means genuine presence becomes rarer.

And when presence becomes rarer, "managed love" can start to look like "good love," simply because it is consistent.

Many people will settle for management because it's predictable.

But predictable is not the same as nourishing.

In the future, as systems become more automated and persuasion becomes cheaper, many relationships will become more performative:

- curated intimacy
- scripted empathy
- outsourced communication
- algorithm-shaped identity

- "connection" as content

So real meeting—unoptimized, unperformed, unedited presence—will become a luxury.

And that is why clarity matters now.

Because if you can't tell the difference between being met and being managed, you will accept a relationship that feels smooth while slowly deleting you.

The Ledger of Love

Clarity in love is not about harsh judgment.

It's about record.

What happens repeatedly?

What do you feel after interactions?

Do you feel expanded—or contracted?

Do you feel more yourself—or less?

Do you feel energized—or drained?

Do you find yourself speaking freely—or choosing your words like you're crossing a minefield?

Love is not proved by occasional grand gestures.

Love is proved by whether your nervous system can relax into truth.

And your nervous system is an honest auditor.

It keeps the ledger even when your mind wants to explain.

Sovereign Love

Sovereign love is not independence-as-avoidance.

It's not "I don't need anyone."

Sovereign love is the ability to be in connection without losing yourself.

It's love that doesn't require self-editing as the entrance fee.

It's love that can hold your becoming.

It's love that can witness your joy without competing with it.

And perhaps most importantly:

It's love that doesn't need you to stay small in order to stay close.

This is the shift clarity creates.

You stop chasing.

You stop proving.

You stop explaining.

You stop auditioning for attention.

You simply begin living as someone who expects to be met.

And if you are not met, you do not argue.

You do not collapse.

You do not rewrite your experience.

You record it.

You accept it.

You choose accordingly.

Clarity in love compacts into a quiet power:

A life that cannot be edited by someone else's discomfort.

A heart that does not negotiate its own truth.

A joy that becomes the test—and the guide.

And once you live that way, something miraculous happens:

Love stops feeling like a room you have to earn your way into.

Love becomes the place where you are already fully allowed.

14.

Clarity in Parenting

Parenting is where clarity stops being inspirational and becomes urgent.

Because in adult life, if people misunderstand you, it hurts—but you can usually recover. You can reframe. You can relocate. You can leave rooms that don't fit.

A child can't.

A child lives inside the room you place them in.

So when I say clarity is an asset, I'm not only talking about money systems or institutions or AI.

I'm talking about the most sacred responsibility I've ever carried:

being a witness for a mind that is still forming.

Especially a gifted mind.

Because giftedness doesn't just mean "smart."

It often means:

- uneven development
- intense curiosity
- unusual sensitivity
- deep moral perception
- rapid leaps followed by plateaus
- boredom with rote tasks
- obsession with certain topics
- language that sounds older than their age
- a nervous system that feels the world more sharply

Giftedness can look like defiance.

It can look like distraction.

It can look like emotional intensity.

It can look like "too much."

And the world has a habit of labeling what it doesn't understand.

THE FIRST BATTLE IS OVER THE STORY

The first "ledger war" a child encounters isn't technological.

It's narrative.

Who gets to tell the story of who they are?

A teacher who sees a behavior and names it a problem.
A system that measures one kind of skill and calls that "ability."
A standardized expectation that punishes divergence.
A well-meaning professional who interprets intensity as pathology.

These aren't villains.

They're machines doing what machines do: categorizing.

But when a child is gifted, categorization can become distortion.

And distortion, repeated, becomes identity.

This is why clarity in parenting matters:

If you don't protect the story, the system will write it for you.

Mislabeling Doesn't Just Misunderstand—It Reshapes

A label can be helpful when it brings support.

But mislabeling has a specific harm: it teaches the child to distrust their own inner reality.

If a child is told, implicitly or explicitly:

- "You're too sensitive."
- "You're dramatic."
- "You're lazy."
- "You're difficult."
- "You're not trying."
- "You're behind."
- "You're disruptive."

…and the child *knows* they are trying, knows they are feeling deeply, knows they are thinking in a different shape—

they begin to split.

They begin to edit themselves to survive the room.

And that self-editing is the beginning of tragedy.

Because once a child learns to abandon their internal witness, they become easy to manage.

They become compliant at the cost of their own mind.

I will not trade my child's coherence for anyone's convenience.

That sentence became one of my deepest parental vows.

Motherhood as Witness

Motherhood—at least for me—became less about control and more about record.

Not a record like surveillance.

A record like truth.

I started paying attention to what my son was actually showing me:

How he thinks.
How he learns.
What lights him up.
What overwhelms him.
What calms him.
How quickly he absorbs complex ideas when he's interested.
How sensitive he is to sound and environment.
How he can move constantly and still be taking everything in.
How he can struggle with a procedural task and then ask a question that feels like philosophy.

The world sees a snapshot and judges.

A mother sees a pattern and holds.

That's witness.

That's the living ledger.

It's the refusal to let a child be reduced to a moment that doesn't represent them.

The school-shaped definition of "smart" is too small

One of the most dangerous cultural edits is the narrowing of intelligence into what can be measured on schedule.

Especially in early childhood.

A child who is brilliant in systems-thinking, language, science curiosity, imagination—can be made to feel "bad at school" because they don't conform to a certain sequence.

A child can be ahead in one domain and behind in another.

A gifted child is often asynchronous.

So the environment matters.

If the environment honors divergence, giftedness becomes confidence.

If the environment punishes divergence, giftedness becomes shame.

I've watched how quickly this can happen.

A child can go from open to guarded in a semester.

Not because they changed.

Because the room changed.

So my job became:

Create rooms where he can stay himself.

And when he must enter rooms that may misunderstand him, give him a compass.

Protecting a Gifted Mind Requires Clarity, Not Defensiveness

This is important: clarity is not fighting everyone.

Clarity is not becoming combative with teachers, tutors, systems.

Clarity is staying anchored to truth without needing to make the world wrong.

It's saying:

- "Here's what I observe over time."
- "Here's what works for him."
- "Here's what shuts him down."
- "Here's where his mind lights up."
- "Here's what he needs to thrive."

It's advocating without panic.

It's refusing to accept mislabeling as "fact."

It's maintaining the child's dignity while navigating the machine.

And as a woman who has spent years inside corporate machines—inside the messy middle of projects where people get misread, undervalued, reduced—I recognized the pattern instantly:

Systems compress reality.

My job is to uncompress my child.

To keep his full humanity intact.

The Difference Between Discipline and Distortion

A gifted child still needs structure.

Still needs boundaries.

Still needs accountability.

But there is a difference between guiding a child and distorting a child.

Guidance says:

- "You can do hard things."
- "You're learning how to persist."
- "We'll find a way that makes sense."
- "Let's build the bridge together."

Distortion says:

- "You are the problem."
- "Your mind is wrong."
- "Your way of learning is unacceptable."
- "You must become smaller to belong."

A gifted child will often comply with distortion because they are perceptive and want to be loved.

That compliance looks like "improvement."

It can also be a slow disappearance.

So I watch for one thing above all:

Does the structure produce more aliveness—or less?

Aliveness is the signal.

PAST VS FUTURE: THE LABELING MACHINE GETS STRONGER

This is not fear mongering. It's just the trajectory.

As education systems incorporate more data, more scoring, more behavioral tracking, more automated assessment, the risk of mislabeling increases—not because the technology is evil, but because measurement can never capture the full interior.

A child becomes a profile.

A set of data points.

A learning "type."

An intervention plan.

And profiles can be helpful.

But profiles can also become cages if they are treated as identity.

In an AI-shaped world, the child who stays coherent—who keeps their own inner witness—will have an advantage that is almost impossible to teach later.

So parenting becomes one of the earliest sites of sovereignty.

If I can help my son keep his internal compass, he will be harder to manipulate in a world of perfect persuasion.

WHAT I'M REALLY PROTECTING

I'm not trying to raise a child who "wins" school.

I'm trying to raise a child who stays himself.

A child who can say:

- "I know what I know."
- "I don't have to pretend."
- "I can learn in my own shape."
- "I can be sensitive and strong."
- "I can be different without being wrong."

That's the deepest form of protection.

Because if he keeps that, he can adapt to any system without being absorbed by it.

He can learn the rules without surrendering his mind.

He can enter institutions without letting institutions write his identity.

MOTHERHOOD AS THE FIRST LEDGER

This is the part that makes me emotional in a way that has nothing to do with sentimentality:

A mother is often the only person who holds the child's full record.

Not the record of grades.

The record of essence.

The record of questions asked in the car.
The record of wonder.
The record of the day they watched rain stop and noticed something no one else noticed.
The record of the strange, brilliant sentence: "Except for the color of nothing."
The record of how their mind moves when they feel safe.

That is not "cute."

That is the raw material of a human being.

And if you don't protect it, the world will compress it into something manageable.

So yes—clarity in parenting is an asset.

It is the willingness to keep the record clean.

To refuse mislabeling.

To advocate without panic.

To build environments where a gifted mind can remain whole.

Because the future will be filled with systems that want to edit people.

And sovereignty begins early, when someone says:

I see you.

I will not reduce you.

I will keep your truth intact until you can keep it yourself.

15.

Clarity Under Pressure

Pressure is the audit.

It reveals what is real—not in the world, but in you.

When life is calm, you can live on good intentions. You can be generous with your interpretations. You can tolerate ambiguity. You can tell yourself stories that keep everything smooth.

But when pressure arrives—money pressure, relationship pressure, institutional pressure, parenting pressure, public pressure—your nervous system stops negotiating.

Under pressure, you don't rise to your ideals.

You return to your defaults.

And that's why this chapter matters: because clarity under pressure isn't a personality trait.

It's a practice.

It's what you do when the world gets loud.

Pressure makes the world persuasive

The first thing pressure does is increase the volume of external signal.

Not just signal from the internet.

Signal from everyone.

People speak with more certainty.
Institutions tighten control.
Rules get enforced harder.
Narratives become more moral and more absolute.
Advice multiplies.
Fear spreads.
Urgency becomes contagious

Pressure makes the environment more persuasive because the human brain is desperate for a fast answer.

Under stress, we want the relief of resolution.

That is exactly when we become easiest to manage.

This is the pattern I've seen professionally and personally:

When people are scared, they reach for the nearest narrative that feels stabilizing.

Not necessarily the truest narrative.

The most stabilizing one.

So pressure doesn't just test character.

It tests discernment.

How I used to handle pressure: I explained and complied

My old default under pressure was the same one I used in relationships and institutions:

I tried to be reasonable.

I tried to find the right words.

I tried to explain myself clearly enough to be understood, approved, allowed.

I tried to reduce friction by absorbing discomfort.

I became cooperative.

This is a common survival strategy, especially for women who have learned that smoothness is safer than conflict.

But it has a hidden cost:

When you use explanation as your pressure response, you lose time.

You lose leverage.

You lose your internal witness.

Because explanation under pressure often isn't communication.

It's negotiation with reality.

And reality doesn't negotiate.

The shift: pressure became my signal to return to record

At some point—after enough messy middles, enough silence from systems, enough moments of being managed—I built a new reflex.

Pressure became my cue to stop talking.

Not forever.

Just first.

First: return to the ledger.

Because the ledger is the one place the world's volume can't reach.

The ledger isn't emotional.

The ledger isn't performative.

The ledger is simply:

What happened.
What repeats.
What was promised.
What was done.
What it cost.
What is true.

When the world gets loud, you don't need more words.

You need reality.

The pressure protocol

This is what clarity under pressure looks like in practice—simple, not glamorous.

1) Reduce input
Under pressure, the first act of sovereignty is narrowing your channels.

Less news. Less scrolling. Less advice. Less argument.

Not forever.

Just long enough to hear your own mind.

2) Name the pressure without narrating it
Pressure makes people dramatize. They turn stress into a story.

Clarity begins with a clean sentence:

I am under pressure right now.

Not: *Everything is falling apart.*

Not: *This means I'm doomed.*

Just: *Pressure is present.*

This keeps you in reality instead of in prophecy.

3) Identify the decision point
Pressure creates false urgency.

So I ask:

What decision must be made—actually?

Not what I *feel* I have to decide.

What do I truly have to decide?

Often the answer is smaller than the panic suggests.

4) Pull the ledger
Now: record.

- What do I know for sure?
- What is pattern vs one-off?
- What are the actual facts?
- What are the incentives of the parties involved?
- What has this cost me before?
- What happens if I do nothing for 24 hours?
- What happens if I act too fast?

This is where clarity returns: facts, patterns, incentives, cost.

5) Separate signal from noise
Under pressure, everything looks like signal.

But much of it is noise:

- other people's fear
- algorithmic amplification
- institutional CYA language
- emotional contagion
- old stories from past wounds
- "helpful" opinions that are really projections

The ledger helps you sort:

Is this evidence—or atmosphere?

6) Choose the anchor
This is Act V in miniature.

Under pressure, you choose what you anchor to:

- my values
- my child's well-being
- my long-term horizon
- my dignity
- verifiable truth
- clean boundaries
- the reality of cost

You can't anchor to everything.

Pressure forces selection.

That's why it's revealing.

7) Act in one clean move
Clarity under pressure often looks boring.

One email.
One boundary.
One payment decision.
One exit.
One refusal.
One request for documentation.
One moment of silence instead of explaining.

Coherence isn't dramatic.

It's precise.

THE INNER PRESSURE: THE TEMPTATION TO REWRITE YOURSELF

There is an external pressure and an internal pressure.

External pressure is obvious: deadlines, demands, conflict, financial stress, institutional threats.

Internal pressure is subtler:

The pressure to revert to your edited self.

To become "nice."
To become "small."
To become "reasonable."
To make everyone comfortable.
To smooth the truth so you don't have to feel the consequence.

That internal pressure is the real battle.

Because if you collapse internally, you can be moved externally.

This is why the ledger matters: it interrupts self-revision.

It holds you to your own truth when your nervous system wants the relief of compliance.

PAST VS FUTURE: PRESSURE WILL BE ENGINEERED

Here is what is changing in the world, and why this chapter is preparation:

In the future, pressure won't only happen.

It will be engineered.

Not always maliciously—just structurally.

Systems will use:

- urgency prompts
- countdown timers
- algorithmic outrage
- "breaking news" cycles
- risk warnings
- compliance deadlines
- identity verification holds
- reputational storms
- synthetic consensus

Pressure will be used as a steering mechanism because it works.

It collapses nuance.

It narrows cognition.

It makes people choose the fastest relief.

So the sovereign person—the coherent person—is the one who can feel pressure and not be steered by it.

THE RETURN

Clarity under pressure is the ability to return.

Return to what you know.

Return to what you've recorded.

Return to what is verifiable.

Return to the cost.

Return to your witness.

Because when the world gets loud, you will be offered infinite words.

And words will feel like answers.

But the ledger is the only thing that holds when the noise is persuasive.

The ledger is what keeps you from being moved by tone.

It keeps you from being managed by narratives.

It keeps you from betraying yourself for relief.

Pressure will come.

It always does.

The question is whether you will meet it as your edited self…

or as your anchored self.

The ledger is how you come back.

Not perfectly.

Not dramatically.

Just consistently.

And consistency is the truest form of power in an era that wants to move you.

16.

The Cost of Delusion

For most of my life, I thought delusion was personal.

A coping strategy.
A blind spot.
A story someone tells themselves because the truth would hurt too much.

And sometimes it is that.

But the longer I lived—especially as the world became more digital, more curated, more optimized—the more I realized something unsettling:

Delusion is no longer only a psychological accident.

Delusion is a product.

It is engineered, packaged, distributed, and monetized.

Not always by villains.

Often by systems doing what systems do: optimizing for engagement, compliance, and profit.

And the human cost is not abstract.

The cost is coherence.

Delusion used to be private. Now it's scalable.

In the past, delusion had limits.

You could live in denial, but reality would still rub up against you in direct ways.

A neighbor might challenge your story.
A family member might confront you.
A community might notice the mismatch.
Consequences arrived with friction.

Now delusion can be supported by infrastructure.

You can find a feed that confirms your story.
A community that validates your avoidance.
A narrative that explains away every contradiction.
A professional vocabulary that turns dysfunction into virtue.
An algorithm that hands you the next reinforcing "proof" before you feel doubt.

So delusion becomes not a glitch in the system.

Delusion becomes a feature.

And features are designed.

HOW DELUSION GETS PACKAGED

Delusion has a certain texture when it's being sold.

It usually comes with:

- certainty without cost
- belonging without accountability
- identity without self-knowledge
- morality without sacrifice
- power without responsibility
- clarity without record

It feels like relief.

It feels like being chosen.

It feels like an answer that finally makes the world make sense.

That's the hook.

Delusion sells because it reduces complexity into a story you can carry.

And human beings love stories that reduce anxiety.

So systems learn to provide them.

DELUSION AS AN EMOTIONAL PRODUCT

This is the part people don't like to admit:

Delusion isn't sold as "falsehood."

It's sold as a feeling.

Safety.
Outrage.
Vindication.
Superiority.
Innocence.
Belonging.

Hope without grief.
Power without vulnerability.

The product isn't the claim.

The product is the emotional state.

That's why arguing facts often fails.

Because the customer didn't buy the delusion for facts.

They bought it for relief.

I RECOGNIZE DELUSION BECAUSE I LIVED THE SMALL VERSION OF IT

Before I saw delusion as a product, I saw it in my own patterns.

I saw how easy it was to rewrite my experience to preserve comfort.

To tell myself:

It's fine.
They didn't mean it.
I'm overreacting.
I should be more reasonable.
This is just how it works.

That wasn't stupidity.

It was strategy.

Because if I admitted what I actually knew, I would have to grieve.

I would have to change.

I would have to confront dynamics that were painful to name.

So I participated in a soft delusion: the delusion of minimizing.

The delusion of making myself smaller so the relationship could remain intact.

The delusion of interpreting management as care.

And the cost showed up as a slow erosion:

Less voice.
Less joy.
Less trust in my own perception.
More fatigue.
More second-guessing.
More internal negotiation.

Delusion's first cost is always internal.

You pay with your witness.

THE MARKET VERSION OF DELUSION IS MORE DANGEROUS

Once you see your personal delusions, you become able to recognize the industrial ones.

Industrial delusion is dangerous for one reason:

It can be mass-produced.

And now—with synthetic media, AI-generated language, algorithmic distribution—it can be personalized.

Delusion can be tailored to your psychology.

To your fears.
To your desires.
To your identity wounds.
To your longing for certainty.
To your hunger for belonging.

This is not science fiction.

This is what optimization does.

Optimization learns what keeps you engaged and gives you more of it.

Sometimes that's education.

Often, it's emotional reinforcement.

So delusion becomes the most profitable content category on earth:

It keeps people watching.

It keeps people clicking.

It keeps people buying.

It keeps people outraged.

It keeps people loyal.

The most expensive delusion is "It will be different next time"

There is one delusion that hides inside almost every human story:

The belief that if we behave better, the system will meet us.

That if we explain better, we will be heard.

That if we comply more fully, we will be safe.

This is the old contract again.

And it is one of the most profitable delusions in modern life, because it keeps people inside managed systems.

It keeps you in the loop:

- submit again
- appeal again
- try again
- wait again
- be patient
- be reasonable

- follow the process
- trust the system

Hope becomes a leash.

And the world offers hope endlessly because hope is cheaper than change.

DELUSION'S TRUE COST: YOU BECOME EDITABLE

This is the theme thread that runs through the whole book.

Delusion makes you editable.

If you can be persuaded to abandon your own witness, you can be rewritten.

By relationships.
By institutions.
By money systems.
By narratives.
By synthetic signal.

Delusion is the mechanism by which power enters the mind without force.

It's not that someone "controls" you in a dramatic way.

It's that you stop insisting on reality.

And when you stop insisting on reality, you become manageable.

That is the cost.

Not that you believe something incorrect.

That you lose the internal structure that keeps you sovereign.

PAST VS FUTURE: DELUSION BECOMES EASIER TO PURCHASE

In the past, delusion required effort.

You had to seek it out. Protect it. Defend it.

In the future, delusion will be delivered like a subscription.

You will be offered:

- AI companions that affirm you
- feeds that mirror you
- realities that never challenge you
- narratives that always make you righteous
- explanations that always soothe you
- communities that reward your most reactive self

This will feel like comfort.

It will also quietly remove the friction that produces growth.

Because growth requires discomfort.

Growth requires honest mirrors.

Growth requires a ledger.

So the future will not only test our politics or our technology.

It will test our capacity to tolerate truth.

THE ANTIDOTE IS NOT CYNICISM

The antidote to delusion is not distrusting everything.

That leads to paralysis.

The antidote is not pride—"I'm too smart to be manipulated."

That is itself a delusion.

The antidote is clarity.

Clarity as orientation.

Clarity as record.

Clarity as the willingness to pay the cost of truth.

Because truth has a cost, too.

Truth costs comfort.
Truth costs belonging sometimes.
Truth costs old stories.
Truth costs fantasies.
Truth costs relationships that depend on your smallness.
Truth costs the soothing relief of "maybe it's fine."

But truth pays you back.

It pays you back in coherence.

And coherence is the new scarcity.

The cleanest way I can say it

Delusion is no longer just what you do to survive.

Delusion is what the world sells you so you don't change.

It is marketed as certainty.

It is delivered as comfort.

It is reinforced as identity.

And the bill always arrives later.

You pay in lost time.
Lost voice.
Lost money.
Lost relationships.
Lost trust in yourself.
Lost ability to tell what is real when the world gets loud.

So this chapter is not a warning.

It's a ledger entry.

A clean, compassionate recognition:

If clarity is the asset, delusion is the debt.

And the interest compounds quietly—until one day you look up and realize you don't recognize your own life.

That is the cost.

And that is why I keep returning to the same practice, again and again:

Return to record.
Return to cost.
Return to what is verifiable.
Return to your witness.

Because in an era where delusion is a product, clarity becomes a form of freedom.

Not because it makes life easy.

Because it makes life real.

The Clarity Aperture

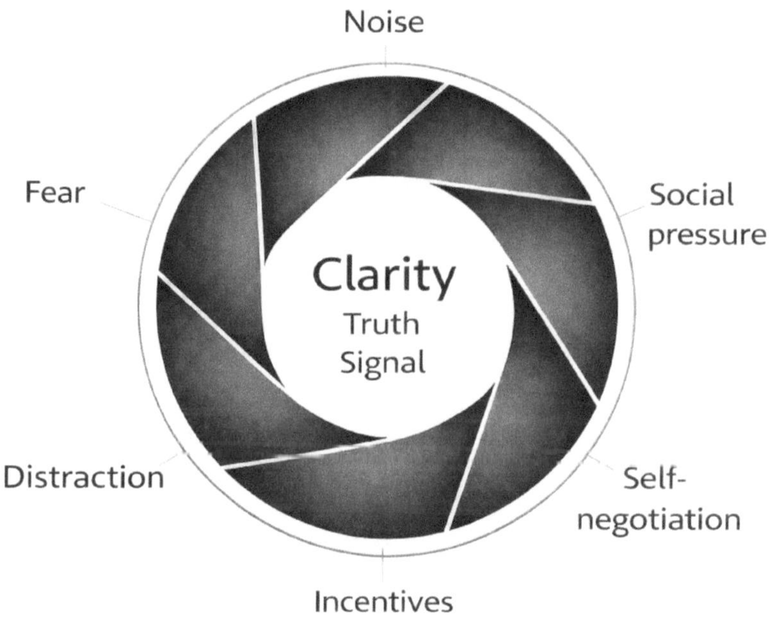

Work is credibility, you can't copy.

17.

The Ledger Principle

The ledger principle is simple enough to sound obvious.

And deep enough to change a life.

What you cannot record, you cannot protect.
What you cannot verify, you will eventually be asked to trust.
What you only "feel" but never name will be rewritten—by others or by you.

A ledger is not just an accounting tool.

A ledger is an architecture of reality.

And the more the world becomes editable—through narrative, through institutions, through AI-generated signal, through invisible data systems—the more this becomes true:

Sovereignty is the ability to keep your reality intact.

That is the ledger principle.

1) Every system is a ledger system

We think "ledger" and we think money.

But money is only the most visible ledger.

The truth is: everything that matters runs on a ledger.

- Relationships keep a ledger of presence, reciprocity, repair.
- Families keep a ledger of belonging, loyalty, history.
- Institutions keep a ledger of eligibility, risk, access.
- Culture keeps a ledger of status, virtue, narrative.
- Technology keeps a ledger of identity, behavior, scoring.
- The self keeps a ledger of truth, grief, boundaries, and what it costs to betray them.

A ledger is simply: *what is counted, what is remembered, what is treated as real.*

And if you don't decide what is counted, something else will.

2) Editability is power

Here is the hidden lever underneath almost every modern conflict:

Who can edit the ledger?

If someone can edit the record, they can shape reality.

They can change what happened.
They can redefine what counts.
They can revise the meaning of your experience.

They can shift the rules without touching the rhetoric.
They can move the goalposts and call it progress.

This is why "being managed" feels like erasure.

Management isn't always loud.

It's often a quiet editing of reality:

- interruptions that break your thread
- "reasonable" rituals that place you under evaluation
- policies that tighten while language softens
- automated decisions with no appeal
- beautiful words with no cost behind them

Editability is the true gate.

And the more the world becomes edit-friendly, the more human beings need anchors.

3) There are two kinds of ledgers: narrative and record

A narrative ledger is based on story.

It can be inspiring.

It can also be manipulated.

Because stories can be rewritten.

A record ledger is based on what holds under scrutiny:

- patterns over time
- verifiable facts
- commitments kept or broken
- costs paid

- consequences observed

The world prefers narrative ledgers because they're flexible.

They keep everything smooth.

But smoothness is not truth.

In an era of synthetic signal, *smoothness is cheap.*

So the ledger principle says: stop building your life on narrative alone.

Build it on record.

4) Your internal ledger is your first defense

This is where the principle becomes intimate.

The first ledger is not a blockchain.

It is your witness.

Your nervous system.
Your memory.
Your body's "yes" and "no."
The quiet part of you that knows when you are being met—and when you are being managed.

Most people lose themselves because they abandon this ledger.

Not once.

Repeatedly.

They explain away the entries.

They call themselves sensitive.

They override their discomfort.

They trade truth for belonging.

They pay for peace with self-erasure.

So the ledger principle begins as a personal vow:

I will not rewrite my own experience to preserve comfort.

That vow is the beginning of coherence.

5) Coherence is proof of work in the self

AI can produce proof of words.

Institutions can produce proof of policy.

Relationships can produce proof of performance.

But only a coherent life produces costly signal.

Coherence is proof of work inside the human being:

- the ability to hold the record
- the ability to act consistently
- the ability to stay aligned under pressure
- the ability to resist being edited by the room

That's why clarity is the first asset.

Because clarity is what makes coherence possible.

And coherence is what remains real when everything else becomes performative.

6) The ledger principle is not paranoia—it's orientation

It's easy to hear "ledger" and think suspicion.

But the ledger principle is not about distrusting everyone.

It's about refusing confusion.

It's about staying literate in a world where:

- words are plentiful
- images are synthesizable
- institutions are automated
- "forms" are invisible
- permission is built into rails
- narratives are optimized
- and delusion is a product

You don't need fear to navigate that world.

You need orientation.

The ledger principle is orientation.

It's the ability to ask, calmly:

- What is being counted here?
- Who can edit it?
- What is the incentive?
- What is verifiable?
- What does it cost?
- What pattern repeats?
- What part of me is being asked to disappear?

Those questions are not political.

They are protective.

7) The capstone choice: what you anchor to

This is the reckoning that becomes sovereignty:

You choose what ledger governs your life.

You choose what you treat as real.

You choose what you will verify.

You choose what you will no longer explain away.

You choose whether you will live in permission… or in record.

And you don't have to become extreme.

You simply become deliberate.

You can use institutions without worshiping them.
You can use rails without surrendering your mind.
You can engage technology without being hypnotized by it.
You can love without editing yourself.
You can parent as witness, not as compliance officer.
You can build a life that can't be rewritten because it is grounded in what holds.

THE LEDGER PRINCIPLE, AS A SINGLE SENTENCE

If I had to distill it to one line—one line I would want my child to carry into the future—it would be this:

When the world becomes editable, anchor to what is verifiable—inside you and outside you—and let your life match your record.

That's it.

Not a manifesto.

A practice.

A way of staying human.

A way of staying coherent.

A way of living so that no system—no relationship, no institution, no narrative, no perfectly generated set of words—can quietly edit you out of your own life.

Because sovereignty is not control over the world.

Sovereignty is the ability to remain yourself inside it.

And the ledger principle is how you do that.

18.

The Last Word

You are entering an era where almost everything can be edited.

Not only photos and videos.

Reality.

The story of what happened.
The story of who you are.
The story of what you meant.
The story of what you owe.
The story of what you said you'd never do again.

The world will offer you endless versions of yourself—polished, reasonable, persuasive, compliant, optimized.

And it will call that progress.

But progress without coherence is just a smoother form of captivity.

So here is my last word:

Do not outsource your witness.

Not to institutions.
Not to crowds.
Not to systems that score you.
Not to feeds that train your attention.
Not to beautiful language—especially not to beautiful language.

Because words are becoming cheap.

Performance is becoming infinite.

And the rarest thing on earth will soon be a human being who can still tell what is real.

Clarity is not a vibe. It's a discipline.

Clarity is not "having the right opinion."

Clarity is not certainty.

Clarity is the courage to hold the record.

To say: *This is what happened.*
To say: *This is what it cost.*
To say: *This is what repeats.*
To say: *This is what I will not explain away anymore.*

Clarity is the refusal to trade your coherence for comfort.

And once you make that refusal, something changes in your life:

You become harder to manage.

YOU WILL BE INVITED INTO DELUSION EVERY DAY

Delusion will come dressed as safety.

As belonging.

As outrage that feels righteous.

As certainty that feels like relief.

As "everyone knows."

As "the experts say."

As "it's probably fine."

As "don't be dramatic."

As "you're overthinking."

As "just trust the process."

Delusion will not announce itself as delusion.

It will announce itself as peace.

But peace that requires you to edit yourself is not peace.

It's compliance.

So when you feel the soft invitation to abandon your own knowing, pause.

Return to the ledger.

Return to what you can verify.

Return to your witness.

KEEP A RECORD THAT CAN'T BE REWRITTEN

This is the simplest form of sovereignty:

Record what matters.

Not obsessively.

Cleanly.

Pattern over time.
Promises and follow-through.
Where you feel met.
Where you feel managed.
Where your joy expands.
Where your joy gets regulated.
Where your body says no long before your mouth does.

A record is not bitterness.

A record is respect for reality.

And reality is the only ground you can build on.

CHOOSE YOUR ANCHORS BEFORE PRESSURE CHOOSES THEM FOR YOU

Pressure is coming.

Not because you are unlucky.

Because pressure is the environment of modern life.

When pressure comes, you will be offered shortcuts:

- a narrative that soothes you
- a leader who sounds certain
- a system that promises safety
- a relationship that demands your smallness
- a version of yourself that makes everyone comfortable

You will be tempted to trade truth for relief.

Don't.

Relief fades.

The bill arrives later.

Choose your anchors now:

- coherence over approval
- record over narrative
- cost over tone
- truth over performance
- dignity over permission

These are not dramatic choices.

They are daily choices.

And daily choices compound.

LET CLARITY COMPOUND INTO A LIFE THAT CAN'T BE EDITED

That is the whole arc of this book.

Clarity is not a single moment.

It is a compounding asset.

Every time you tell the truth to yourself, you compound it.
Every time you keep your boundary, you compound it.
Every time you stop explaining, you compound it.
Every time you choose what is verifiable, you compound it.
Every time you stay coherent under pressure, you compound it.

And eventually, almost without realizing it, you become someone whose life cannot be rewritten by someone else's discomfort.

You become stable in your own reality.

Not rigid.

Stable.

This is what sovereignty actually is.

Not control over others.

Control over your own edits.

MY FINAL CHARGE

Do not let the future steal your mind with perfect words.

Do not let convenience become your compass.

Do not let managed systems convince you that your discomfort is a defect.

Do not let your joy be regulated by people who cannot hold your becoming.

Do not let the world's noise replace your inner record.

Instead:

Be the person who keeps the ledger clean.
Be the person who can verify.
Be the person who chooses costly signal over cheap persuasion.
Be the person who stays coherent when the room gets loud.

Because coherence will be the new wealth.

And the people who have it will be the ones who can still love without disappearing, parent without mislabeling, build without performing, and walk through an editable world without becoming editable themselves.

This is not a political stance.

It's orientation.

It is how you stay human.

It is how you stay free.

And it begins the moment you decide:

I will not abandon my witness.

Tech Interlude E

What To Do With This

If you've read this far, you might be waiting for the part where I give you the steps.

The ten rules.
The morning routine.
The "do these five things and you'll be sovereign."

I understand that impulse. It's human. We want closure. We want a handle. We want to take a complicated world and turn it into something we can manage.

But the whole point of this book is that we're leaving the era where life can be managed by simple instructions.

We're entering an era where the most dangerous thing is *borrowed certainty*—certainty that feels good, sounds clean, and requires no internal work.

So what I want to offer here is not a checklist.

It's a stance.

A way of standing in the world that makes you harder to edit.

A way of moving through a persuasion-rich, data-driven, increasingly automated environment without losing your witness.

You can apply this stance to money. To institutions. To parenting. To love. To technology. To your own mind.

And if you hold it long enough, it begins to compound.

1) Keep your ledger

Not because you're suspicious.

Because you're alive.

A ledger is not paranoia. It's memory with integrity.

A ledger is the refusal to let your reality be rewritten by:

- someone else's tone
- a system's opacity
- a perfectly phrased explanation
- your own desire to avoid discomfort
- the social pressure to be "reasonable"

Keeping your ledger means you don't have to argue about reality.

You don't have to beg to be understood.

You don't have to rehearse the same story until it feels acceptable.

You simply stay in contact with what happened.

What repeats.

What it costs.

The most important ledger isn't even written down—though sometimes writing it down helps.

The most important ledger is internal:

- What makes you expand?
- What makes you contract?
- Where do you feel met?
- Where do you feel managed?
- What do you consistently explain away?
- What do you consistently regret after you comply?

Your nervous system is an honest auditor.

It keeps the record even when your mouth performs.

So the stance begins with a simple refusal:

I will not rewrite my own experience to preserve comfort.

That sentence will change your life if you mean it.

2) Choose dependencies on purpose

You will depend on systems.

Anyone who tells you otherwise is selling a fantasy.

You will use banks and platforms and schools and services and devices and institutions, because modern life is built out of dependencies.

The question is not: *Will I depend?*

The question is: *What am I willing to depend on—and what do I refuse to make essential?*

This is where most people get trapped: they drift into dependencies unconsciously.

They build their life on rails they don't understand, then they're shocked when the rails behave like rails:

- conditional access
- policy changes
- scoring
- holds
- opaque decisions
- "for your safety" language that feels like a door closing

A sovereign stance doesn't demonize rails.

It simply sees them.

It asks:

- Where is the gate?
- Who can edit the rules?
- What is the incentive?
- What happens under pressure?
- What happens if I'm misclassified?
- What happens if the platform changes terms?
- What happens if the institution optimizes against me?

Then you decide what you will *not* make single-point-of-failure.

You diversify dependencies the way a wise person diversifies risk:

Not with panic.

With clarity.

You don't need to burn your life down.

You just stop giving any one system the power to define your reality.

3) Protect attention like it's a bank account

Because it is.

Attention is the currency of the next era.

And most people are going to be poor—not in money, but in mind.

They will live inside constant input:

- outrage cycles
- urgency cues
- algorithmic persuasion
- synthetic consensus
- perfect "clarity" delivered as content
- communities that reward reactivity
- endless commentary that feels like knowing

The result isn't education.

The result is a mind that can't hold a thought.

A mind that can't feel the difference between signal and atmosphere.

A mind that can't build.

A mind that is permanently steerable.

So the stance is not "avoid the internet."

The stance is:

My attention is not an open port.

I decide what enters.
I decide when it enters.
I decide what gets repeated inside me.

This is not puritanical.

It's survival.

In a world where persuasion is cheap, your attention becomes expensive.

And if you don't protect it, you will slowly lose the ability to tell what you actually believe.

You'll start mistaking emotional activation for truth.

You'll start mistaking familiarity for reality.

You'll start thinking you're informed when you're simply saturated.

Protecting attention isn't about being strict.

It's about being sovereign.

4) Verify before believing

Not everything.

Just what matters.

This is where nuance lives.

The point isn't to become cynical.

The point is to stop using surface polish as proof.

AI can produce perfect words.

Institutions can produce perfectly reasonable explanations.

A crowd can produce perfectly unanimous consensus.

None of that guarantees truth.

So the stance becomes:

- Don't let language be evidence.
- Don't let tone be proof.
- Don't let consensus be certainty.

Instead, look for costly signal:

- consistency over time
- accountability under pressure
- pattern that persists when it's inconvenient
- verifiable record
- willingness to be audited
- "skin in the game" behavior rather than performance

This is why proof of work matters.

AI can produce signal.

Only work produces costly signal.

And costly signal is what holds when the environment becomes synthetic.

Verification isn't a personality.

It's humility.

It's the recognition that the world can be persuasive and you can be moved.

So you slow down.

You ask: *What do I actually know?*

You separate:

- evidence
  from

- interpretation
  from
- emotion

This is not coldness.

It's maturity.

5) Become literate in editability

This is the quiet skill that will define the next decade:

Knowing what can be edited—and by whom.

In money systems, you ask:

- Who can freeze it?
- Who can inflate it?
- Who can reverse it?
- Who can surveil it?

In identity systems, you ask:

- Who decides I'm "verified"?
- Who decides I'm "risky"?
- What happens if I'm misclassified?

In relationships, you ask:

- Who edits reality through interruption?
- Who edits joy through regulation?
- Who edits the past when confronted?

In institutions, you ask:

- What do they optimize for?

- What do they avoid at all costs?
- What can never be provided here?

Editability is power.

And literacy in editability is freedom.

Not because you escape all systems.

Because you stop being surprised by them.

And being unsurprised is a kind of peace.

6) Choose the kind of person you will be in an editable world

This is the deepest part of the stance.

Because you can protect your attention and diversify dependencies and verify what matters—and still be lost if you don't decide what kind of inner life you're building.

In an editable world, many people will become editable too.

They'll become:

- personality-shaped by incentives
- mood-shaped by feeds
- values-shaped by the room
- identity-shaped by approval
- beliefs-shaped by social pressure
- memory-shaped by narrative

The stance I'm pointing to is not hard.

It's steady.

It's saying:

I will be coherent.

Not perfect.

Coherent.

I will live in a way that my future self will not have to explain.

I will not trade my witness for belonging.

I will not mistake performance for love.

I will not outsource reality to systems that cannot meet me.

I will let clarity be the asset that compounds.

That's sovereignty.

Not as politics.

As orientation.

THE SIMPLEST WAY TO HOLD IT

If you want one sentence to carry forward, it's this:

Keep your ledger, choose your dependencies, protect your attention, and verify before you believe—so your life can't be edited by the loudest system in the room.

That's not a program.

It's a stance.

And if you hold it—gently, consistently—you won't need to be afraid of the future.

Because you won't be drifting inside it.

You'll be oriented.

You'll be anchored.

You'll be building a life that can take pressure without losing its shape.

A life that still belongs to you.

Appendix A

The Proof-of-Work Toolkit

Proof of work matters, not words.

In an era where language is cheap, endlessly editable, and increasingly synthetic, the only thing that holds weight is what can be demonstrated. In Bitcoin, proof of work secures reality. In human life, proof of work does the same thing: it keeps you anchored to what is true.

This appendix is not theory. It is a set of ledger tools—simple tests that help you stop arguing with narratives and start reading the actual record.

A1. Proof of Work vs. Proof of Words

Proof of Words is what someone says they are.

It is identity-by-declaration: intentions, promises, credentials, sincerity, branding, "I'm the kind of person who…," "Trust me."

Proof of Work is what someone does—repeatedly—when it costs them something.

It is character-by-evidence: patterns, receipts, follow-through, repair, restraint, consistency across time.

In an editable world, words can be rearranged forever.

Work leaves a trail.

The simplest test:

- If it's real, it appears in behavior.
- If it's performance, it appears in speech.

A2. What Counts as Proof

In relationships

- Consistency over intensity.
- Repair after harm, not persuasion after harm.
- Mutuality: effort flows both ways without leverage.
- Respect for autonomy: no management, no "correcting," no coercion.
- Truth without punishment: you can speak freely without consequences.

In leadership and institutions

- Transparency: what is documented matches what is said.
- Accountability: consequences exist, not just statements.
- Systems that work: measurable outcomes, not mission language.
- Ethics under pressure: behavior doesn't change when nobody is watching.
- Consistency across people: rules apply to insiders too.

In money and incentives

- Aligned incentives: the model doesn't require deception to survive.
- Skin in the game: risk is shared, not exported to others.
- Auditability: records can be checked.
- Durability: value holds beyond the pitch.

A3. Proof-of-Words Red Flags

You don't need to prove intent. You only need to recognize structure.

- Performed humility: humility that is really a hook for admiration.
- Endless explaining: long narratives meant to override your perception.
- Selective disclosure: keeping you uncertain to keep you available.
- Authority posture: scolding, correcting, diagnosing, ranking, managing.
- Charm as leverage: warmth that always steers you somewhere.
- Credentials as a substitute for evidence: "I'm qualified" replacing "I delivered."
- Promises without receipts: future talk that never becomes a record.
- You feel smaller after contact: less clear, more fog, more self-doubt.

A simple rule:

If you leave a conversation more confused than when you entered, you were likely being managed.

A4. The Ledger Questions

Use these when you feel yourself slipping into confusion, rationalization, or hope.

1. What is the pattern—not the exception?
2. What do they do when they don't get what they want?
3. What does this relationship require me to ignore?
4. Do I feel met—or managed?
5. Do they respect my "no" without punishment?
6. Do their words predict their behavior?
7. What does it cost them to treat me well? (If it costs nothing, it's not evidence.)
8. Is the effort mutual—or am I carrying the emotional labor?
9. If I stop explaining, what remains?
10. If I imagine this continuing for five years, do I feel expanded—or reduced?

Ledger rule:

If you need to keep persuading yourself, something is off.

A5. Boundary Scripts

You do not owe debates. You do not owe explanations.

A boundary is not an argument; it is a decision.

When someone tries to manage you

- "I'm not available for being managed."
- "I'm not looking for feedback on my decisions."
- "That doesn't work for me."
- "Let's change the subject."
- "I'm going to end this conversation now."

When someone sells empowerment but builds dependence

- "I'm focused on what reduces dependency."
- "I don't need this."
- "If it's real, it will show up in outcomes."
- "I'm not interested in ongoing engagement."

When someone wants access without mutuality

- "I'm keeping my life private."
- "I'm not available for favors."
- "I'm not the right person for that."
- "Wishing you well."

Use the calm repeat:

Say it once. If they push, repeat it verbatim. Then exit.

A6. The Proof-of-Work Standard

Proof of work matters, not words.

And the companion line:

In an editable world, proof is the only audit trail.

When you live by this, you stop negotiating with performance.

You stop arguing with narratives.

You stop needing permission.

You simply read the ledger—and act accordingly.

Choose what's true.
Pay the cost. Keep the signal.

About the Author

Lisa Scott is the CEO and founder of In2Edge, where she helps private equity and corporate teams navigate the "messy middle" after the deal—those high-stakes transition and integration seasons where reality matters more than presentations. With a career built at the intersection of operations, contracts, and execution, she's known for turning complexity into clear systems that hold under pressure.

Lisa's work has long centered on truth in motion: what can be verified, what can be measured, and what must be owned when the room gets loud. While she isn't a technologist by training, she has spent years translating real-world business friction into practical tools, platforms, and repeatable operating models—building ledgers that keep teams aligned when narratives shift.

A Standard Partner of the Texas Blockchain Council, Lisa is part of a growing community exploring how proof, transparency, and decentralized infrastructure can strengthen human systems—money, institutions, and everyday life. Her writing blends lived experience with a clear-eyed view of the future, offering readers a grounded way to stay coherent in an increasingly editable world.

Lisa is also the creator of the podcast *M&A+ The Art After the Deal* and the founder of Lisa J. Scott Publishing. She lives in the Dallas area with her children and writes at the intersection of sovereignty, systems, and the quiet power of clarity.

Books by Lisa J. Scott

The Core Trilogy

The Woman from Frog Creek

A memoir of origin, sovereignty, and lived refusal.

Following the Compass

On movement, orientation, and learning to trust direction.

The Fixed Point

Stillness, identity, and what remains when motion stops.

Learning & Formation

The Color of Nothing

Lessons on learning, attention, and the space between.

Outside the Bell

A book about children whose intelligence does not ring on cue.

The Bird Who Looked First

A story about early seeing.

The Frog Creek School

The lived environment where learning happens.

Work, Systems & Aftermath

The Art After the Deal

What remains when the transaction is over.

When the Forms Stopped Coming

Short stories from a post–income-tax world.

When Money Changed Quietly

On invisible systems and everyday life.

Meetings I Survived

Notes on sanity under bureaucratic absurdity.

Relational & Perceptual Works

Rooms of Friendship

On the architecture of relationship.

Borrowed Certainty

On identity formed through inherited belief.

The Mirror Held Twice

Perception, reflection, and seeing again.

The Second Axis: The Mirror, Held Again

On perception after reflection.

The Bright Part

Attention, gentleness, and what draws the eye.

Clarity, Sovereignty & Signal

Clarity

On clarity as the most valuable asset.

Capstone

The Wayfinder Maps

An integration of the whole.

LISA J. SCOTT
PUBLISHING